Palgrave Studies in Digital Business & Enabling Technologies

Series Editors
Theo Lynn
Irish Institute of Digital Business
DCU Business School
Dublin, Ireland

Pierangelo Rosati
J.E. Cairnes School of Business and Economics
University of Galway
Galway, Ireland

This multi-disciplinary series will provide a comprehensive and coherent account of cloud computing, social media, mobile, big data, and other enabling technologies that are transforming how society operates and how people interact with each other. Each publication in the series will focus on a discrete but critical topic within business and computer science, covering existing research alongside cutting edge ideas. Volumes will be written by field experts on topics such as cloud migration, measuring the business value of the cloud, trust and data protection, fintech, and the Internet of Things. Each book has global reach and is relevant to faculty, researchers and students in digital business and computer science with an interest in the decisions and enabling technologies shaping society. More information about this series at http://www.palgrave.com/gp/series/16004.

Theo Lynn
Pierangelo Rosati
Mohamad Kassem
Stelios Krinidis • Jennifer Kennedy
Editors

Disrupting Buildings

Digitalisation and the Transformation of Deep
Renovation

Editors
Theo Lynn
Irish Institute of Digital Business
DCU Business School
Dublin City University
Dublin, Ireland

Mohamad Kassem
School of Engineering
Newcastle University
Newcastle upon Tyne, UK

Jennifer Kennedy
Irish Institute of Digital Business
DCU Business School
Dublin City University
Dublin, Ireland

Pierangelo Rosati
J.E. Cairnes School of Business and
Economics
University of Galway
Galway, Ireland

Stelios Krinidis
Information Technologies Institute
Centre for Research & Technology
Hellas (CERTH)
Thessaloniki, Greece

Department of Management Science
and Technology
International Hellenic University
Kavala, Greece

ISSN 2662-1282 ISSN 2662-1290 (electronic)
Palgrave Studies in Digital Business & Enabling Technologies
ISBN 978-3-031-32308-9 ISBN 978-3-031-32309-6 (eBook)
https://doi.org/10.1007/978-3-031-32309-6

This Palgrave Macmillan imprint is published by the registered company Springer Nature Switzerland AG.
The registered company address is: Gewerbestrasse 11, 6330 Cham, Switzerland

Acknowledgements

This book was partially funded by the European Union's Horizon 2020 Research and Innovation Programme through the RINNO project (https://rinno-h2020.eu/) under Grant Agreement 892071, and the Irish Institute of Digital Business.

Book Description

The world's extant building stock accounts for a significant portion of worldwide energy consumption and greenhouse gas emissions. In 2020, buildings and construction accounted for 36% of global final energy consumption and 37% of energy-related CO_2 emissions. The European Union (EU) estimates that up to 75% of the EU's existing building stock has poor energy performance, 85–95% of which will still be in use in 2050.

To meet the goals of the Paris Agreement on Climate Change will require a transformation of construction processes and deep renovation of the extant building stock. The World Economic Forum, World Business Council for Sustainable Development, and the European Commission are amongst the many global organisations that recognise the important role ICTs can play in construction, renovation, and maintenance, as well as supporting the incentivisation and financing of deep renovation. Technologies such as sensors, big data analytics and machine learning, building information modelling (BIM), digital twinning, simulation, robots, cobots and unmanned autonomous vehicles (UAVs), additive manufacturing, smart contracts, and the Internet of Things are transforming the deep renovation process, improving sustainability performance, and developing new services and markets.

This book defines a deep renovation digital ecosystem for the twenty-first century, providing a state-of-the art review of current literature, suggesting avenues for new research, and offering perspectives from business, technology, and industry.

CONTENTS

Notes on Contributors

Mazen J. Al-Kheetan is an Assistant Professor in the Department of Civil and Environmental Engineering at Mutah University, Jordan. He is also an associate editor at the Proceedings of the Institution of Civil Engineers—Transport, UK. He previously served as the head of the Department of Civil and Environmental Engineering at Mutah University, Jordan.

Marco Arnesano is Associate Professor of Mechanical and Thermal Measurements and coordinator of Industrial Engineering at eCampus University, Italy. He is also the co-founder of LIS (Live Information System), a startup company developing BIM-based solutions for buildings' digitalisation.

Ioannis Brilakis is Laing O'Rourke Professor of Construction Engineering and the director of the Construction Information Technology Laboratory in the Department of Engineering at the Division of Civil Engineering, University of Cambridge, UK.

Mehdi Chougan is a Marie Skłodowska-Curie Research Fellow in the Department of Civil and Environmental Engineering at Brunel University London, UK. His research focuses on cementitious composite materials, especially in graphene-engineered cementitious composites, and additive manufacturing of alkali-activated cementitious composites.

Mark Cummins is Professor of Financial Technology at the University of Strathclyde, UK. His research interests include financial technology

(FinTech), quantitative finance, energy and commodity finance, sustainable finance, and model risk management.

Borja García de Soto is Assistant Professor of Civil and Urban Engineering at New York University Abu Dhabi (NYUAD), UAE, and a Global Network Assistant Professor in the Department of Civil and Urban Engineering at the Tandon School of Engineering, New York University (NYU), USA. He is the director of the S.M.A.R.T. Construction Research Group at NYUAD and his research focuses in the areas of automation and robotics in construction, cybersecurity in the AEC industry, artificial intelligence, lean construction, and BIM.

Asimina Dimara is a Research Assistant at Centre for Research & Technology Hellas/Information Technologies Institute (CERTH/ITI), Greece. She holds an MSc by research in Intelligent Computer Systems from the University of the Aegean and is currently undertaking a PhD in Intelligent Computer Systems from the University of the Aegean.

Omar Doukari is Research Fellow in Construction Informatics at the University of Northumbria Newcastle, UK. He completed his PhD in Computer Science and Artificial Intelligence, focusing on spatial information representation, modelling, and reasoning.

Antonia Egli is the Communication & Dissemination Manager for the H2020-funded deep renovation project, RINNO, and Research Fellow at the Irish Institute of Digital Business and *safe*food. As a postgraduate researcher, she focuses on identity, stigma, and the spread and influence of misinformation within the vaccine discourse on social media.

Seyed Hamidreza Ghaffar is Professor of Civil Engineering. He is a Chartered Civil Engineer (CEng, MICE), a Member of the Institute of Concrete Technology (MICT), and a Fellow of Higher Education Academy (FHEA). He is the founder and director of Additive Manufacturing Technology in Construction (AMTC) Research Centre at Brunel University London, UK.

David Greenwood is Professor of Construction Management at Northumbria University, UK, and director of BIM Academy. He has written widely and delivered consultancy around the world.

Zhiqi Hu is an early stage researcher under the Fellowship Marie Skłodowska-Curie Actions and a PhD candidate in the Department of

Engineering at the Construction Information Technology Laboratory, University of Cambridge, UK.

Dimosthenis Ioannidis is a Senior Researcher at the Information Technologies Institute of the Centre for Research & Technology Hellas and Lecturer in Interdepartmental Postgraduate Programs at the Aristotle University of Thessaloniki, Greece.

Mohamad Kassem is Professor of Digital Construction Management at Newcastle University, UK. He has established expertise in development and application of digital and data-centric tools and methods in construction and infrastructure management. In this domain, he authored over 130 papers and successfully secured and completed multi-million-pound research and innovation grants.

Jennifer Kennedy is a Postdoctoral Researcher at the Irish Institute of Digital Business at DCU Business School. Dr Kennedy specialises in knowledge processes with a specific focus on how tacit knowledge is transferred between novice and experts in the workplace.

Paraskevas Koukaras is a Postdoctoral Researcher at the Information Technologies Institute of the Centre for Research & Technology Hellas and Lecturer in Postgraduate Programs at the School of Science and Technology, International Hellenic University, Greece.

Stelios Krinidis is an Assistant Professor in the Department of Management Science & Technology at the International Hellenic University, Greece, and a postdoctoral researcher at the Information Technologies Institute of the Centre for Research & Technology Hellas.

Theo Lynn is Full Professor of Digital Business at Dublin City University, Ireland, and co-director of the Irish Institute of Digital Business. He was formerly the principal investigator (PI) of the Irish Centre for Cloud Computing and Commerce, and director of the LINK Research Centre. Lynn specialises in the role of digital technologies in transforming business processes and society with a specific focus on cloud computing, social media, and data science.

Silvia Angela Mansi is a PhD student in Science Applied to Wellness and Sustainability at eCampus University, Italy. Her research activity is focused on sensors for indoor comfort measurement with multi-domain approaches.

Yuandong Pan is a PhD student at the Chair of Computational Modeling and Simulation and Institute for Advanced Study, Technical University of Munich, Germany.

Alessandro Pracucci is Innovation Manager at Focchi Group, Italy. He has expertise in multidisciplinary research particularly in technological, digital, and sustainability relations. He leads the Department of Innovation at Focchi Group, investigating new growth opportunities for the company.

SeyedReza RazaviAlavi is an assistant professor at Northumbria University, UK. Prior to joining Northumbria University, he was a postdoctoral fellow in the Construction Simulation research group at the University of Alberta, and worked five years as a project management consultant in Canada.

Pierangelo Rosati is Associate Professor of Digital Business and Society at University of Galway, Ireland. His research interests include digital business, business value of IT, FinTech, blockchain, cloud computing, and cyber security.

Muammer Semih Sonkor is a Research Assistant at New York University Abu Dhabi (NYUAD), UAE, and a PhD student at New York University (NYU), USA. He worked on several large-scale construction projects before attending the European Master's Program in Building Information Modeling (BIM A+). He conducts research focusing on cybersecurity in construction as a part of the S.M.A.R.T. Construction Research Group at NYUAD.

Christos Tjortjis is the Dean of the School of Science and Technology at the International Hellenic University, Greece, director for 5 MSc programmes, and Associate Professor of Knowledge Discovery and Software Engineering Systems.

Dimitrios Tzovaras is a Senior Researcher and the president of the Board of the Centre for Research & Technology Hellas, Greece.

Laura Vandi is a Project Manager in the Department of Innovation at Focchi Group, Italy. She manages and develops internal and European projects regarding sustainability in the construction sector with particular focus on retrofitting, technologies, and circular economy. She spent years abroad and worked in architectural firms.

ABBREVIATIONS

4M   Mapping, Modelling, Making and Monitoring
API   Application Programming Interface
AR   Augmented Reality
AWS   Amazon Web Services
B2B   Business-to-Business
B2B2C   Business-to-Business-to-Consumer
B2C   Business-to-Consumer
BACS   Building Automation and Control Systems
BEMS   Building Energy Management Systems
BIM   Building Information Modelling
CIB   International Council for Research and Innovation in Building and
          Construction
CO_2 Carbon Dioxide
CSP   Cloud Service Provider
EC   European Commission
EU   European Union
GDP   Gross Domestic Product
GHG   Greenhouse Gas
HVAC   Heating, Ventilation, Air Conditioning
IAAS   Infrastructure as a Service
ICT   Information and Communications Technologies
IDDS   Integrated Design and Delivery Solutions
IEEE   Institute of Electrical and Electronics Engineers
IPCC   Intergovernmental Panel on Climate Change
ISO   International Organisation for Standardisation
ISV   Independent Software Vendor
KPI   Key Performance Indicator

kWh	Kilowatt Hour
m^2	Metres Squared
NZEB	Near Zero Energy Building
OSS	One Stop Shop
PV	Photovoltaic
QoE	Quality of Experience
QoS	Quality of Service
RES	Renewable Energy Source
SDG	Sustainable Development Goal
UN	United Nations
VR	Virtual Reality

LIST OF FIGURES

LIST OF TABLES

CHAPTER 1

Deep Renovation: Definitions, Drivers and Barriers

Theo Lynn, Pierangelo Rosati, and Antonia Egli

Abstract This chapter defines the key elements of the deep renovation life cycle. Investment in deep renovation is driven by various rationales, including societal, economic, environmental, energy security, quality, opportunistic, and catalytic motivations and benefits. At the same time, both deep renovation and digital technology adoption to support deep renovation are impacted by challenges presented in humans, organisational processes, technologies and external environments. This chapter explores the key drivers and barriers to deep renovation and associated digitalisation. It establishes the motivation for the remainder of the book.

T. Lynn (✉) • A. Egli
Irish Institute of Digital Business, DCU Business School,
Dublin City University, Dublin, Ireland
e-mail: theo.lynn@dcu.ie; antonia.egli@dcu.ie

P. Rosati
J.E. Cairnes School of Business and Economics,
University of Galway, Galway, Ireland
e-mail: pierangelo.rosati@universityofgalway.ie

1

Keywords Deep renovation • Energy efficiency • Residential buildings • Renovation life cycle • Adoption • Barriers to adoption

1.1   INTRODUCTION

The 2022 Intergovernmental Panel on Climate Change (IPCC) assessment suggests that climate-resilient development is already challenging at current warming levels and that the window for action to address climate change is narrowing. Restricting warming to around 2°C (3.6°F) still requires global greenhouse gas (GHG) emissions to peak before 2025 at the latest and be reduced by a quarter by 2030 (IPCC, 2022). This is a significant challenge. In the words of UN Secretary-General António Guterres, "the climate emergency is a race we are losing, but it is a race we can win" (UN, 2019).

The European Union (EU) is not sitting idle. As part of the European Green Deal, the EU has raised its ambition to reduce GHG emissions by 2030 from its previous target of 40% to at least 55% below 1990 levels, as well as increasing the share of renewable energy by 32%, and improving energy efficiency by 32.5% (European Commission, 2022). The renovation of EU building stock is particularly critical in supporting these goals. Buildings are among the most significant sources of energy use within the EU: existing structures currently account for 40% of all energy consumption and 36% of GHG emissions (European Commission, 2020a). In particular, it is estimated that over 75% of the EU's residential building stock has poor energy performance, the majority of which will be still in use by 2050 (European Commission, 2021a). To meet its climate change goals, the EU seeks to achieve a decarbonised EU building stock by 2050. To achieve this, it has recently put in place measures to consolidate its existing goals, encourage the use of digital technologies and smart applications in building operations, and strengthen the links between achieving higher renovation rates, funding and energy performance certification (European Commission, 2021a). Deep renovation is key to achieving this goal.

The remainder of this chapter will explore narrow and broad definitions of deep renovation including the rationales for undertaking deep renovation. Recent research by Lynn et al. (2021) suggests such rationales not merely are related to environmental sustainability but include a wide range of different stakeholder motivations including economic, energy security and opportunistic rationales, amongst others. Notwithstanding these rationales, the widespread deep renovation of building stock, particularly in a constrained time frame, faces significant barriers not least human, organisational, technological and environment context challenges. We discuss how these barriers may surface across the life cycle of a deep renovation project. Advances in technologies, not least information and communications technologies (ICTs), are central to accelerating the renovation life cycle and overcoming the existing barriers to deep renovation. We conclude with a summary of the remainder of this book which looks at the main digital innovations disrupting and transforming the construction sector.

1.2 DEEP RENOVATION

"Deep renovation" has become somewhat of a buzzword in recent years, albeit an obscure one. There remains little consensus on the term's definition and, although widely adopted in academia, industry and legislation, definitions vary significantly on local, regional and international levels (Shnapp et al., 2013). While deep renovation (sometimes referred to as deep energy renovation, deep retrofit or deep refurbishment) may be defined simply as renovation efforts which capture the "full economic energy efficiency potential of improvement works […] of existing buildings" and lead to high energy performance levels (Shnapp et al., 2013), the core concept of deep renovation is categorised into *broad* and *narrow* definitions:

- *Broad*, referring to the use of different simultaneous building envelope and installation system renovation measures into one integrated strategy across the entire building life cycle (Agliardi et al., 2018);
- *Narrow*, relating to performance levels of refurbishments that reduce building energy consumption by a significant proportion to energy levels observed before renovation works began (Sibileau et al., 2021).

D'agostino et al. (2017) take a more quantitative approach, categorising deep renovation efforts by performance impact as presented in Table 1.1. This offers a relative numeric classification of deep renovation efforts, although an exact quantitative reference value for deep renovation energy reductions remains unavailable (D'Oca et al., 2018).

Deep renovation involves the use of multiple energy-saving measures. Bruel et al. (2013) summarise these measures as (1) energy-efficient building elements such as windows, heating, ventilation and air conditioning (HVAC), air filtration, lighting and appliances; (2) renewable energy sources like solar hot water, solar photovoltaic (PV) panels, passive solar energy, shading, wind, heat pumps, biomass and biogas; and (3)

Table 1.1   Categorisations of deep renovation measures

Deep renovation class	Description
Minor	Reduces final energy consumption levels by up to 30% by implementing one to three improvement measures and costing an average of 60 €/m2
Moderate	Involves more than three improvements to existing buildings resulting in energy reductions of 30–60%
Deep	Enabled through high-grade improvements that result in energy savings ranging between 60% and 90% and costing between 140 and 330 €/m2
Major	Covers renovation works on more than 25% of the building envelope and costs more than 25% of the value of the existing building
Nearly Zero Energy Buildings (NZEP)	Results in buildings which perform significantly better in energy use (with +90% of improved final energy saving) and rely on renewable energy sources (RES) ideally produced within or near the building itself

community energy sources such as district heating systems. Each of these measures alone improves energy performance in buildings and may be employed in combination with traditional technology and construction solutions (D'Oca et al., 2018). However, deep renovation is distinct from other energy-efficient retrofits in that these elements become *fully integrated* within the renovation process.

1.3   RATIONALES AND BENEFITS OF DEEP RENOVATION

In 2018, EU renovation rates barely exceeded 1% and were significantly below the objectives set in the Energy Efficiency Directive (Directive 2012/27/EU) and the revised Energy Performance Building Directive (Directive 2018/844). Only 11% of the EU building stock undergoes renovation on a yearly basis (European Commission, 2021a). Reaching the 2030 and 2050 goals requires a significant acceleration and greater understanding of what drives stakeholders to adopt and implement a deep renovation strategy. An attempt at this is made in Table 1.2.

Aside from advancing building quality and area net-worth in comparison to other buildings through state-of-the-art aesthetic, safe and easy-to-use building elements, deep renovation reinforces economic stimuli in the form of employment and reduced reliance on international energy imports

Table 1.2   Stakeholder rationales towards adopting deep renovation practices

Stakeholder group	Description	Rationale towards adopting deep renovation practices
Energy solutions and construction technology providers, and independent software vendors (ISVs)	Develop and market (1) technologies that support and deliver residential deep renovation projects and (2) software solutions for building information modelling, deep renovation process management, building and infrastructure management and maintenance, and/or related technology management	Aim to improve, extend or complement existing product or service offerings in a cost-effective way, increase competitiveness through value-added products and services, and generate incremental revenues with comparatively little upfront R&D investment
Housing development and construction companies	Buy, license and use technologies and systems developed by third parties to deliver high energy performance	Aim to differentiate themselves from competitors by providing superior services and buildings and generating more profit from these projects while delivering better performance and value for their clients
Architects	Design and plan the renovation and construction of built environments while using a variety of software tools and related databases to model and design built environment projects	Require specific skills, tools and knowledge for gathering environmental and cultural considerations, both pre- and post-occupancy, as well as implementing specific sustainable designs and smart technologies
Construction finance companies and crowdfunding platforms	Provide capital to construction companies for financing projects, possibly from alternative sources of capital such as crowdfunding platforms	Traditional sources of capital cannot fully meet the financial demand of deep renovation works and are oftentimes constrained by regulation. Deep renovation crowdfunding may offer faster, transparent and more secure options for all stakeholders

(continued)

Table 1.2 (continued)

Stakeholder group	Description	Rationale towards adopting deep renovation practices
Building owners	Own and manage residential buildings	Aim to renovate their building stock cost efficiently while minimising disturbance to occupants and overall renovation time
		Aim to increase energy efficiency and environmental performance to meet or exceed national standards, meet domestic or European policy goals, maximise occupant satisfaction and ultimately increase the value of the property
Occupants	Reside in the building in question, often during the course of renovation works	Require renovation solutions which offer the best value for money in the form of long-term energy performance. Sometimes, reducing environmental impact and meeting or exceeding international standards for energy performance is a priority
Research centres and projects	Attract government and industry funding to carry out research on existing renovation solutions, the economic and business impacts of novel solutions, or the industry adoption of novel technologies and processes	Focus on specific elements of the (deep) renovation life cycle as a research field and, in doing so, operate within pre-defined boundaries and aim to influence a large number of stakeholders
Investors and licensors	Invest or license technology and other research outputs for commercial purposes	Aim to differentiate themselves from competitors by providing superior, better-performing technologies to their clients

(*continued*)

Table 1.2 (continued)

Stakeholder group	Description	Rationale towards adopting deep renovation practices
EU institutions, policymakers, and funding and standardisation bodies	Formulate or influence policy in EU institutions and national and local government and include regulators, international bodies and other political bodies	Are driven by set national and international climate targets and, in an effort to reach these targets, regulate and secure funding for the advancement of deep renovation projects
Media and industry analysts	Create content to influence stakeholders and possibly perform primary and secondary market research within an industry such as information technology and telecommunications	Disseminate content surrounding the latest developments and technologies in renovation and construction industries with a growing interest in sustainable and energy-efficient practices

(Jochem & Madlener, 2003; Baek & Park, 2012; Bruel et al., 2013; Ferreira & Almeida, 2015; D'Oca et al., 2018; European Commission, 2020). Currently, approximately 34 million Europeans are impacted by energy poverty or the inability to afford adequate heating or lighting (European Commission, 2020b). As such, deep renovation supports citizens in participating in a greener society first-hand while simultaneously improving energy security, health and accessibility for society's most vulnerable citizens (Baek & Park, 2012; Bruel et al., 2013; Ferreira & Almeida, 2015; European Commission, 2020). Deep renovation works lastly deliver improved consumer service on public, community and commercial levels (Jochem & Madlener, 2003; Baek & Park, 2012; Guerra-Santin et al., 2017; Klumbyte et al., 2020).

If properly integrated, deep renovation efforts create resilient and green living spaces while promoting high energy performance and lower waste and pollution levels (Baek & Park, 2012; Bruel et al., 2013; Ferreira & Almeida, 2015; Haase et al., 2020). From a wider perspective, such efforts lead to improved quality of life for building occupants, increased revenues and decreased technological and operational costs through superior products and services, improved security, quality and control over full project life cycles, and more durable buildings in the long term (Mainali et al.,

2021). Perhaps most importantly, deep renovation may positively influence public attitudes towards climate change mitigation works, substitute existing, climate-damaging methods in the traditionally conservative construction sector and improve the uptake of novel and existing ClimateTech and CleanTech measures (Baek & Park, 2012; Mainali et al., 2021).

1.4   BARRIERS TO DEEP RENOVATION

Prior literature presents an extensive range of theoretical lenses by which to explore technology adoption and use, typically from an adopter-centred or innovation or organisation-centred perspective. These lenses are summarised in Table 1.3.

Table 1.3   Theoretical overview of technology adoption and use

Perspective	Theory	Description	Source(s)
Adopter-centred	Theory of Reasoned Action (TRA)	Posits that human behaviour is determined by intention, which in turn is influenced by attitude (towards the behaviour) and subjective social norms (e.g., normative beliefs, demographic variables and personality traits)	Fishbein and Ajzen (1977)
Adopter-centred	Theory of Planned Behaviour (TPB)	An extension of TRA in that it includes the element of perceived behavioural control, that is, facilitating or impeding factors which influence the performance of a behaviour in question	Ajzen (1991)
Adopter-centred	Technology Acceptance Model (TAM)	Reflects elements of TRA, but specifically concerns levels of acceptance across end-user computing technologies including perceived usefulness and perceived ease of use	Davis (1985, 1989)
Adopter-centred	Unified Theory of Technology Acceptance and Use (UTAUT)	Defines determinants of user acceptance and usage behaviour based on performance expectancy, effort expectancy, social influence and facilitating conditions	Venkatesh et al. (2003); Venkatesh et al. (2012)

(*continued*)

Table 1.3 (continued)

Perspective	Theory	Description	Source(s)
Innovation or organisation-centred	Diffusion of Innovation (DOI)	Includes constructs like relative advantage, ease of use, image, visibility, compatibility, results demonstrability and voluntariness of use to define individual technology acceptance	Rogers (1995, 2003)
Innovation or organisation-centred	Technology-Organisation-Environment (TOE) Framework	Identifies the degree to which technological, organisational and environmental aspects influence the process of adopting and implementing a technological innovation	Tornatzky and Fleischer (1990)
Innovation or organisation-centred	Human-Organisation-Technology Fit (HOT-fit)	Used to evaluate information systems based on human (i.e., user satisfaction and system use), technological (i.e., system, information and service quality) and organisational (i.e., environment and structure) dimensions	Yusof et al. (2008)

Although the individual arguments for shortcomings in acceptance towards deep renovation measures lie beyond the scope of this chapter, it is worth noting that the success of deep renovation efforts is impacted by adopter-centred factors, technology-related factors, organisational factors and external environmental factors. The following sections elaborate on these potential reasons for failure.

1.4.1 Human Barriers to Deep Renovation Adoption and Use

Barriers to accepting, supporting and adopting climate-friendly technologies and practices in buildings are manifold. Hesitancy can be traced back to restrictive social norms and household characteristics, short-termism and lack of clarity surrounding the negative consequences of climate change, as well as inadequate knowledge or reservations about the existence or use of new technologies (Van Raaij & Verhallen, 1983; Curtis et al., 1984; Scott, 1997; Abrahamse et al., 2005; Organisation for Economic Co-operation and Development, 2011; Mills & Schleich, 2012;

Huebner et al., 2013; Giraudet, 2020). Demographics such as age, education, household composition and geographical location have equally been shown to affect the adoption of energy efficiency technologies. For example, Mills and Schleich (2012) find that families with young children (unlike elderly household members) are more likely to adopt energy-efficient technologies, as are those with higher education levels. Interestingly, data suggests a high degree of country heterogeneity with respect to adoption, use and attitudes towards household energy-efficient technologies and energy conservation practices (Mills & Schleich, 2012).

In their ethnographic study of the occupants and users of a multi-dwelling residential building in Italy, Prati et al. (2020) find that enhanced quality of life and long-term financial savings were the primary motivators for accepting and supporting deep renovation projects for tenants. However, the economic burden does not fall on tenants, suggesting a need for a *multi-stakeholder approach* to deep renovation projects particularly where there is a divergence in ownership and occupancy. While levels of normative legitimacy may be relatively high amongst tenants (considering the largely accepted moral obligation of preserving the environment), pragmatic legitimacy may be restrained by conflicts between building owners' self-interest, perceived utility, and financial and time requirements of renovation works. Research suggests that barriers influencing the self-interest and utility involved in deep renovation measures include occupant disturbance and a lack of awareness, understanding and trust in deep renovation and new technologies (D'Oca et al., 2018; Prati et al., 2020; European Commission, 2020). Further individual adopter factors include performance expectancy, effort expectancy and social influence (Fishbein & Ajzen, 1977; Ajzen, 1991; Davis, 1985, 1989; Venkatesh et al., 2003, 2012).

Psychological (and oftentimes geographical) distance to the climate crisis is a key barrier amongst consumers in mitigating the effects of climate change and maintaining pro-environmental behaviours (Spence et al., 2012). In one scenario, this may result in building owners and occupants failing to adopt energy management measures in an individual building and within the context of that building's location and climate. As a consequence, this usually leads to unnecessarily high energy and emissions levels (Jochem & Madlener, 2003). In this context, *short-termism* has had a particularly negative impact on the adoption of deep renovation projects. For example, there is a substantial literature base which acknowledges that the adoption of energy-efficient measures is related to cost (Curtis et al.,

1984; Abrahamse et al., 2005; Organisation for Economic Co-operation and Development, 2011). Consumers are more likely to adopt low-cost or no-cost measures much unlike deep renovation projects. As Mills and Schleich (2012) note, such behavioural changes may only have transitory effects, while energy savings resulting from technology adoption tend to have more long-term effects. Consequently, although the adoption of energy-efficient technologies can have a significant impact on the wider environment, it does not necessarily compensate energy savers and thus presents a significant challenge in persuading the public to act (Mills & Schleich, 2012). Particularly in the context of multi-dwelling residential buildings, this may cause mismatches between individual needs and beliefs and those of the wider collective (D'Oca et al., 2018).

Notably, *solution aversion* to climate-friendly measures, which occurs when problems are ignored due to dissatisfaction with proposed solutions, may also impact openness towards deep renovation efforts (Campbell & Kay, 2014). *Tangible solution aversion* in particular applies to the deep renovation context. Poortinga et al. (2004), for example, warn that environmental attitudes may be too limited in explaining environmental behaviour and related technology adoption—particularly because addressing climate change results in tangible lifestyle changes for building occupants. For this reason, deep renovation solutions must pay attention to promoting the cost of non-action and life-quality benefits in ways that can be received by different audiences in different climate and building-type contexts.

1.4.2 Technological Barriers to Deep Renovation Adoption and Use

Technology-related adoption factors include, amongst others, innovation characteristics, availability, ease of use, compatibility, results demonstrability and quality-driven factors (Tornatzky & Fleischer, 1990; Rogers, 1995, 2003; Yusof et al., 2008). Key focus points over previous years have shifted from the technical suitability of deep renovation technologies primarily to the *integration* of energy-saving technologies throughout deep renovation projects (D'Oca et al., 2018). This includes building envelopes, HVAC systems and RES-powered systems (D'Oca et al., 2018). Today, the main technological challenge to deep renovation lies in the complexity associated with integrating technically viable, context-appropriate technologies according to desired outcomes and regulatory standards (Attia et al., 2017). Because of this, one could posit that meeting standards of deep renovation requirements, for example, the Passive House Standard, is less a matter of the technological state of the art, but rather technical awareness, availability and know-how (Innovate UK, 2013; De Gaetani et al., 2020). In its worst case, a lack thereof can lead to missed opportunities, inadequate performance and dissatisfaction with deep renovation as a concept.

The issue of integrating technologies into the building renovation process becomes particularly complex when one considers the abundance of domains, stakeholders and outbound dependencies to systems, regulations and geographical characteristics related to the deep renovation process. This is an issue of *interoperability*, that is, "the ability of two or more systems or components to exchange information and to use the information that has been exchanged" (ISO, 2013). The definition of interoperability has morphed somewhat over time, initially used to describe "a feature of information systems that enabled information exchange" to any system which is able to collaborate with another system (Turk, 2020). Its value becomes evident in enabled communication, coordination, cooperation, collaboration and distribution (Grilo & Jardim-Goncalves, 2010). Unfortunately, the range of heterogeneous applications and systems used by different stakeholders varies across the project life cycle. An example of this is Building Information Modelling (BIM), which presents a plethora of varying software tools designed for energy simulation, planning and management (El Asmi et al., 2015; Arayici et al., 2018). Lacking interoperability (particularly when combined with the dynamic nature of construction projects) becomes an issue in that data flows and value generation are negatively affected by data mismatches, data quality issues and inconsistent sector standards and processes (Curry et al., 2013; Arayici et al., 2018; Shirowzhan et al., 2020). While interoperability with other systems, for example, Geographic Information System (GIS) and Augmented Reality (AR)/Virtual Reality (VR), has been increasingly prioritised, knowledge and practice gaps for integrating stateof-art technologies remain (Shirowzhan et al., 2020).

This is not to say that the technological status quo does not face quality or performance issues in itself (Attia et al., 2017). Primarily the adoption and use of software- or cloud-enabled solutions is inflicted by poor on-site connectivity and latency, lack of integration across supply chains, inconsistent data flows and inadequate worker skills (Almaatouk et al., 2016; Bello et al., 2020). A further by-product of the Internet of Things or smart or otherwise connected products is copious volumes of data—all originating from end points with varying capabilities, connectivity levels, requirements and priorities. Due to the idiosyncrasies of individual buildings and living spaces, owners and occupiers, and the environment in which they are located, this requires taking into account both local and more global considerations (Venkatesh, 2008).

1.4.3   Organisational Barriers to Deep Renovation Adoption and Use

Organisation-related barriers to deep renovation include organisation size and structure, adequacy of resources, top management support and perceived indirect benefits (Tornatzky & Fleischer, 1990; Rogers, 1995, 2003; Yusof et al., 2008). Because of the multi-stakeholder nature of deep renovation projects, existing resources, technical competencies and innovation levels amongst management and operational teams vary and must be considered (Yusof et al., 2008). Resource allocation, financial investment and employee competency all have the potential to hinder deep renovation uptake. For example, research finds that inadequately trained professionals and construction workers within the realm of energy efficiency present a significant barrier to project success (Innovate UK, 2013; Attia et al., 2017; D'Oca et al., 2018; Vavallo et al., 2019).

From an organisational perspective, financial barriers are amongst the most highly cited in literature (Cooremans & Schönenberger, 2019; Bertoldi et al., 2021). This is accelerated by the complexity of deep renovation, particularly in multi-residential buildings such as social housing or other fragmented ownership models (D'Oca et al., 2018). Procurement policies which prioritise price over the quality of renovations, combined with high upfront investment costs and challenging access to funding, may negatively affect deep renovation efforts initiated by the public sector (European Commission, 2017; Van Oorschot et al., 2019; D'Oca et al., 2018; European Commission, 2020). In its worst case, this can result in project delays, underwhelming energy performance and heightened costs—finally leading to reduced consumer trust in public sector efforts overall and specifically deep renovation projects (D'Oca et al., 2018).

1.4.4   *External Environment Barriers to Deep Renovation Adoption and Use*

The external environment, including building and environmental regulations, policies and standards, heavily impacts deep renovation. Environmental factors encompass all external pressures on deep renovation initiatives, including regulatory, competitive and financial pressures, as well as related support from public bodies and partners (Tornatzky & Fleischer, 1990). For those involved in the supply chain, keeping up with changing regulatory requirements can be a significant challenge particularly under changing political administrations.

Legislation and regulation are highlighted as potentially obtrusive to deep renovation efforts in that these are often complex, unclear and time-consuming (European Commission, 2017; D'Oca et al., 2018; European Commission, 2020). One reason for this is that the context of local governments, and more specifically local energy issues, is often ignored in EU regulations or other intergovernmental treaties. Here, central governments are mainly targeted and expected to oversee the implementation of climate objectives (European Commission, 2017). Because they are responsible for the implementation of energy-saving measures, local entities have specific insights into the barriers they face and must therefore become more closely involved in the development of deep renovation strategies, regulations and targets (European Commission, 2017). In a cross-European report, main local barriers to deep renovation were found to be primarily fiscal and financial (i.e., referring to lack of technical skills for funding applications, poorly designed or lack of incentives, limited borrowing capacity, complex financial schemes and unfavourable accounting rules), followed by legislative and strategic barriers such as an incomplete overview of building stocks, limited training in deep renovation practices and lack of technical capacity required for such projects (European Commission, 2017). As previously identified in Sect. 1.2, one final clear strategic barrier was deemed to be the lack of a uniform definition of deep renovation itself (European Commission, 2017).

1.5   Conclusion

This chapter introduces deep renovation, which involves renovation works that capture the full potential of energy- and cost-saving adjustments to existing buildings, along with its benefits and the human, technological, organisational and external environment barriers associated with deep renovation projects. Deep renovation has the potential to transform the construction and renovation industry in its integrated use of multiple energy-saving measures. Projects simultaneously offer relief for vulnerable residential consumer groups, further desperately needed climate-friendly and potentially net-zero energy practices, and heighten the long-term durability of buildings. Each chapter of this book is dedicated to exploring the impact of a specific digital technology on the implementation and delivery of deep renovation projects. Chapter 2 is dedicated to embedded sensors, one of the (if not the most important) enabling technology in the digitalisation of deep renovation. In fact, the use of sensor networks and connectivity represents a key prerequisite for measuring, and therefore optimising, the energy performance of an existing building and for efficient construction management. This chapter presents the role of sensor networks in the field of deep renovation, introduces the concept of smart buildings and smart homes and their main advantages and benefits, and highlights some of the main challenges and concerns associated with the use of sensor infrastructures which are mostly related to the volume, access and use of data captured by sensors on an ongoing basis.

Chapter 3 focuses on BIM, which leverages the large volume of data generated by sensor networks to manage "[…] the information on a project throughout its whole life cycle" (Hamil, 2022). Chapter 3 explores the evolution of BIM from its emergence in the early 1990s to recent developments and describes different BIM "dimensions". The chapter continues by presenting how BIM enables multi-criteria decision-making in the context of building renovation, and deep renovation more specifically, and how it can help to identify, optimise, validate and communicate different renovation scenarios and corresponding costs, timelines and effectiveness. The chapter concludes with a discussion of two main sets of barriers to BIM adoption, namely interoperability and the lack of ontologies that are specifically designed for renovation work which undermine the potential for process automation.

Another way of leveraging the vast amount of data generated by sensors is to develop models that evaluate the energy performance of an existing building and estimate how changes in external and internal conditions would affect such a performance. This technique is called Building Performance Simulation (BPS) and is the main topic of Chap. 4. More specifically, this chapter provides an overview of the main approaches and applications of BPS in the context of deep renovation and discusses how to integrate simulations with real-time monitoring and diagnostic systems for building energy management and control.

Chapter 5 is dedicated to the application of Big Data and analytics in the deep renovation with a particular focus on Machine Learning and Artificial Intelligence and the changes they have enabled in the various phases of the renovation project life cycle, from the renovation design to post-renovation monitoring and assessment. The chapter presents a series of use cases and applications of Big Data in construction and discusses the main advantages and benefits (e.g., alternative design automation, the development of accurate performance prediction models, higher efficiency and reduced environmental impact of the renovation work), as well as the main barriers and challenges (human, technological and organisational) to the wider adoption of Big Data and analytics in deep renovation.

When it comes to capturing data about the physical structure of an existing building, detailed information can be gathered by adopting 3D scanning tools and techniques which enable the creation of a digital twin of the building. Chapter 6 introduces this novel technological paradigm in more detail, describes the main steps and approaches to creating digital twins and presents three main use cases for digital twins in the built environment, namely condition monitoring, facility management and environment simulation. The chapter concludes with a discussion of the main challenges associated with adopting and using digital twins which are mostly related to the high cost and effort required to create the digital twin.

Chapters 7 and 8 turn the attention to the construction phase of the renovation life cycle. In fact, Chap. 7 focuses on additive manufacturing (often referred to as 3D printing) which is the process of fabricating three-dimensional objects following a specific computer design. Additive manufacturing has attracted growing attention from the construction sector in recent years as it promises lower waste and costs, and it provides the opportunity to create complicated large-scale structures and integrate functional building elements such as pipes and storage units within the structure itself. These benefits are discussed in more detail alongside some practical challenges (e.g., equipment costs, skills and lack of standardisation) that are adversely impacting the diffusion of this technology.

Chapter 8 focuses on the use of intelligent equipment and robots (IER) in construction sites. This chapter discusses the maturity of IER technologies that are currently available in the market, describes how they can be used both on-site (e.g., inspection, construction and maintenance) and off-site (e.g., factories) and discusses the key concerns and barriers to adoption which are mostly related to high costs, lack of skills, human-robot interactions and security.

The issue of security is not only relevant in the context of IER, but it is a recurring concern across the entire renovation life cycle. This topic is discussed in more depth in Chap. 9 which provides an overview of relevant cybersecurity frameworks, standards, guidelines and codes of practice. These include, for example, relevant International Organization for Standardization (ISO) and American Institute of Certified Public Accountants (AICPA) standards, the NIST Framework for Improving Critical Infrastructure Cybersecurity, and the European Union Network and Information Systems (NIS) Directive. The chapter concludes by highlighting the need for a contingency approach to assess and manage cyber risk in the context of building renovation, as a one-size-fits-all approach may not be desirable or feasible given the variety of stakeholders involved in this kind of projects.

The final chapter discusses how novel financial technology (fintech) solutions such as crowdfunding, peer-to-peer lending and blockchain-based mechanisms such as tokenisation can help building owners and construction companies overcome one of the main barriers to deep renovation, access to capital. The chapter outlines the main advantages and benefits of these alternative sources of finance, as well as the challenges associated with each of these funding mechanisms, and concludes with a call for further research on both demand side (fund seekers) and supply side (investors) incentives and dynamics or indeed on the responsibilities of platforms that enable and facilitate these transactions.

REFERENCES

Abrahamse, W., Steg, L., Vlek, C., & Rothengatter, T. (2005). A review of intervention studies aimed at household energy conservation. *Journal of Environmental Psychology, 25*(3), 273–291.

Agliardi, E., Cattani, E., & Ferrante, A. (2018). Deep energy renovation strategies: A real option approach for add-ons in a social housing case study. *Energy and Buildings, 161*, 1–9.

Ajzen, I. (1991). The theory of planned behaviour. *Organizational Behavior and Human Decision Processes, 50*(2), 179–211.

Almaatouk, O., Othman, M. S. B., & Al-Khazraji, A. (2016). A review on the potential of cloud-based collaboration in construction industry. In *2016 3rd MEC International Conference on Big Data and Smart City (ICBDSC)* (pp. 1–5). IEEE.

Arayici, Y., Fernando, T., Munoz, V., & Bassanino, M. (2018). Interoperability specification development for integrated BIM use in performance based design. *Automation in Construction, 85,* 167–181.

Attia, S., Eleftheriou, P., Xeni, F., Morlot, R., Ménézo, C., Kostopoulos, V., Betsi, M., Kalaitzoglou, I., Pagliano, L., & Cellura, M. (2017). Overview and future challenges of nearly zero energy buildings (nZEB) design in Southern Europe. *Energy and Buildings, 155,* 439–458.

Baek, C. H., & Park, S. H. (2012). Changes in renovation policies in the era of sustainability. *Energy and Buildings, 31047,* 485–496.

Bello, S. A., Oyedele, L. O., Akinade, O. O., Bilal, M., Delgado, J. M. D., Akanbi, L. A., Ajayi, A. O., & Owolabi, H. A. (2020). Cloud computing in construction industry: Use cases, benefits and challenges. *Automation in Construction.*

Bertoldi, P., Economidou, M., Palermo, V., Boza-Kiss, B., & Todeschi, V. (2021). How to finance energy renovation of residential buildings: Review of current and emerging financing instruments in the EU. *Wiley Interdisciplinary Reviews: Energy and Environment, 10,* e384.389.

Bruel, R., Fong, P., & Lees, E. (2013). *A guide to developing strategies for building energy renovation.* Buildings Performance Institute Europe. [online]. Retrieved July 2022, from https://www.worldgbc.org/sites/default/files/Developing-Building-Renovation-Strategies.pdf

Campbell, T. H., & Kay, A. C. (2014). Solution aversion: On the relation between ideology and motivated disbelief. *Journal of Personality and Social Psychology, 107*(5), 809.

Cooremans, C., & Schönenberger, A. (2019). Energy management: A key driver of energy-efficiency investment? *Journal of Cleaner Production, 230,* 264–275.

Curry, E., O'Donnell, J., Corry, E., Hasan, S., Keane, M., & O'Riain, S. (2013). Linking building data into cloud: Integrating cross-domain building data using linked data. *Advanced Engineering Informatics, 27,* 206–219.

Curtis, F. A., Simpson-Housley, P., & Drever, S. (1984). Household energy conservation. *Energy Policy, 12*(4).

D'agostino, D., Zangheri, P., & Castellazzi, L. (2017). Towards nearly zero energy buildings in Europe: A focus on retrofit in non-residential buildings. *Energies, 10*(1), 117.

D'Oca, S., Ferrante, A., Ferrer, C., Pernetti, R., Gralka, A., Sebastian, R., & op 't Veld, P. (2018). Technical, financial, and social barriers and challenges in deep building renovation: Integration of lessons learned from the H2020 cluster projects. *Buildings, 8*(12), 174.

Davis, F. D. (1985). *A technology acceptance model for empirically testing new end-user information systems: Theory and results.* Doctoral dissertation, Massachusetts Institute of Technology.

Davis, F. D. (1989). Perceived usefulness, perceived ease of use, and user acceptance of information technology. *Management Information Systems Quarterly, 13*, 319–340.

De Gaetani, C. I., Mert, M., & Migliaccio, F. (2020). Interoperability analyses of BIM platforms for construction management. *Applied Sciences, 10*, 4437.

El Asmi, E., Robert, S., Haas, B., & Zreik, K. (2015). A standardised approach to BIM and energy simulation connection. *International Journal of Design Sciences and Technology, 21*, 59–82.

European Commission. (2017). Barriers that hinder deep renovation in the building sector. [online]. Retrieved April 2022, from https://ec.europa.eu/research/participants/documents/downloadPublic?documentIds=080166e5b3a088c0&appId=PPGMS

European Commission. (2020). A renovation wave for Europe—Greening our buildings, creating jobs, improving lives. [online]. Retrieved July 2022, from https://ec.europa.eu/energy/sites/ener/files/eu_renovation_wave_strategy.pdf

European Commission. (2020a). Impact assessment—Stepping up Europe's 2030 climate ambition: Investing in a climate-neutral future for the benefit of our people. [online]. Retrieved April 2021, from https://eur-lex.europa.eu/legal-content/EN/TXT/289/HTML/?uri=CELEX:52020SC0176&from=EN

European Commission. (2020b). Questions and answers on the renovation wave. [online]. Retrieved October 2021, from https://ec.europa.eu/commission/presscorner/detail/en/qanda_20_1836

European Commission. (2021a). Preliminary analysis of the long-term renovation strategies of 13 Member States. [online]. Retrieved April 2021, from https://ec.europa.eu/energy/sites/default/files/swd_commission_preliminary_analysis_of_member_state_ltrss.pdf

European Commission. (2021b). 2030 climate & energy framework. [online]. Retrieved April 2021, from https://ec.europa.eu/clima/policies/284strategies/2030_en

European Commission. (2022). 2030 climate & energy framework. [online]. Retrieved April 2021, from https://ec.europa.eu/clima/eu-action/climate-strategies-targets/2030-climate-energy-framework_en#greenhouse-gas-emissions%2D%2D-raising-the-ambition

Ferreira, M., & Almeida, M. (2015). Benefits from energy related building renovation beyond costs, energy and emissions. *Energy Procedia, 78*, 2397–2402.

Fishbein, M., & Ajzen, I. (1977). Belief, attitude, intention, and behaviour: An introduction to theory and research. *Philosophy and Rhetoric, 10*(2).

Giraudet, L. G. (2020). Energy efficiency as a credence good: A review of informational barriers to energy savings in the building sector. *Energy Economics, 87.*

Grilo, A., & Jardim-Goncalves, R. (2010). Value proposition on interoperability of BIM and collaborative working environments. *Automation in Construction, 19,* 522–530.

Guerra-Santin, O., Boess, S., Konstantinou, T., Herrera, N. R., Klein, T., & Silvester, S. (2017). Designing for residents: Building monitoring and co-creation in social housing renovation in the Netherlands. *Energy Research & Social Science, 32,* 164–179.

Haase, M., Lolli, N., & Thunshelle, K. (2020). *Renovation concepts for residential buildings: Research status, challenges and opportunities.* ZEN Report.

Hamil, S. (2022). *What is BIM?* NBS Enterprises. Retrieved April 30, 2022, from https://www.thenbs.com/knowledge/what-is-building-information-modelling-bim

Huebner, G. M., Cooper, J., & Jones, K. (2013). Domestic energy consumption—What role do comfort, habit, and knowledge about the heating system play? *Energy and Buildings, 66,* 626–636.

Innovate UK. (2013). Retrofit revealed—The Retrofit for the future project data analysis report. [online]. Retrieved August 2022, from https://www.gov.uk/government/publications/retrofit-revealed-retrofit-for-the-future-data-analysis

IPCC. (2022). Sixth assessment report. [online]. Retrieved August 2022, from https://www.ipcc.ch/report/ar6/wg2/resources/press/press-release/

ISO. (2013). Information and documentation—Thesauri and interoperability with other vocabularies—Part 2: Interoperability with other vocabularies. [online]. Retrieved July 2022, from https://www.iso.org/cms/render/live/en/sites/isoorg/contents/data/standard/05/36/53658.html

Jochem, E., & Madlener, R. (2003). The forgotten benefits of climate change mitigation: Innovation, technological leapfrogging, employment, and sustainable development. In *Workshop on the benefits of climate policy: Improving information for policy makers* (p. 318). OECD Paris.

Klumbyte, E., Bliudzius, R., & Fokaides, P. (2020). Development and application of municipal residential buildings facilities management model. *Sustainable Cities and Society, 52.*

Lynn, T., Rosati, P., Egli, A., Krinidis, S., Angelakoglou, K., Sougkakis, V., Tzovaras, D., Kassem, M., Greenwood, D., & Doukari, O. (2021). RINNO: Towards an open renovation platform for integrated design and delivery of deep renovation projects. *Sustainability, 13*(11), 6018.

Mainali, B., Mahapatra, K., & Pardalis, G. (2021). Strategies for deep renovation market of detached houses. *Renewable and Sustainable Energy Reviews, 138,* 110659.

Mills, B., & Schleich, J. (2012). Residential energy-efficient technology adoption, energy conservation, knowledge, and attitudes: An analysis of European countries. *Energy Policy, 49*, 616–628.

Organisation for Economic Co-operation and Development. (2011). *Greening household behaviour: The role of public policy.* OECD Publishing.

Poortinga, W., Steg, L., & Vlek, C. (2004). Values, environmental concern, and environmental behaviour: A study into household energy use. *Environment and Behavior, 36*, 70–93.

Prati, D., Spiazzi, S., Cerinšek, G., & Ferrante, A. (2020). A user-oriented ethnographic approach to energy renovation projects in multi apartment buildings. *Sustainability, 12*(19), 8179. https://doi.org/10.3390/su12198179

Rogers, E. (1995). *Diffusion of innovations* (4th ed., pp. 15–23). The Free Press.

Rogers, E. (2003). *Diffusion of innovations* (p. 337). Simon and Schuster.

Scott, S. (1997). Household energy efficiency in Ireland: A replication study of ownership of energy saving items. *Energy Economics, 19*, 187–208.

Shirowzhan, S., Sepasgozar, S. M., Edwards, D. J., Li, H., & Wang, C. (2020). BIM compatibility and its differentiation with interoperability challenges as an innovation factor. *Automation in Construction, 112*, 103086.

Shnapp, S., Sitha, R., & Laustsen, J. (2013). What is deep renovation? [online]. Retrieved April 2021, from https://www.gbpn.org/sites/default/files/304/08.DR_TechRep.low_pdf

Sibileau, H., Broer, R., Dravecký, L., Fabbri, M., Álvarez, X. F., Kockat, J., & Jankovic, I. (2021). Deep renovation: Shifting from exception to standard practice in EU policy. [online]. Retrieved July 2022, from https://www.bpie.eu/wp-content/uploads/2021/11/BPIE_Deep-Renovation-Briefing_Final.pdf

Spence, A., Poortinga, W., & Pidgeon, N. (2012). The psychological distance of climate change. *Risk Analysis: An International Journal, 32*(6), 957–972.

Tornatzky, L., & Fleischer, M. (1990). *The process of technology innovation.* Lexington Books.

Turk, Z. (2020). Interoperability in construction—Mission impossible? *Developments in the Build Environment, 4*, 100018.

UN. (2019). Remarks at 2019 climate action summit. [online]. Retrieved August 2022, from https://www.un.org/sg/en/content/sg/speeches/2019-09-23/remarks-2019-climate-action-summit

Van Oorschot, J., Halman, J., & Hofman, E. (2019). The continued adoption of housing systems in the Netherlands: A multiple case study. *Journal of Construction Engineering, Management & Innovation, 2*, 167–190.

Van Raaij, W. F., & Verhallen, T. M. (1983). A behavioural model of residential energy use. *Journal of Economic Psychology, 3*, 39–63.

Vavallo, M., Arnesano, M., Revel, G. M., Mediavilla, A., Sistiaga, A. F., Pracucci, A., Magnani, S., & Casadei, O. (2019). Accelerating energy renovation solution for zero energy buildings and neighbourhoods—The experience of the RenoZEB project. *Multidisciplinary Digital Publishing Institute Proceedings, 20*, 1.

Venkatesh, A. (2008). Digital home technologies and transformation of households. *Information Systems Frontiers, 10*(4), 391–395.

Venkatesh, V., Morris, M. G., Davis, G. B., & Davis, F. D. (2003). User acceptance of information technology: Toward a unified view. *MIS Quarterly, 27*, 425–478.

Venkatesh, V., Thong, J. Y., & Xu, X. (2012). Consumer acceptance and use of information technology: Extending the unified theory of acceptance and use of technology. *MIS Quarterly, 36*, 157–178.

Yusof, M. M., Papazafeiropoulou, A., Paul, R. J., & Stergioulas, L. K. (2008). Investigating evaluation frameworks for health information systems. *International Journal of Medical Informatics, 77*(6), 377–385.

CHAPTER 2

Embedded Sensors, Ubiquitous Connectivity and Tracking

Marco Arnesano and Silvia Angela Mansi

Abstract The digitalisation of the deep renovation process and built environment is enabled by ubiquitous connectivity and monitoring of the environment itself, the artefacts and actors within it, and events that occur. Such monitoring is important for efficient construction management, dynamic peak demand reduction, affordability, and occupants' well-being. Sensor networks based on Internet of Things (IoT) technologies represent an important prerequisite for both optimising and redefining the stages of the building process to meet environmental challenges. This chapter provides an overview of how computation capabilities are being integrated into the physical environment and the role of sensor networks in the context of deep renovation. The key advantages and benefits of these technologies at the pre, during and post-renovation stages are discussed together with different use cases. The value of sensor network infrastructures and the legal and ethical implications of the use of such sensor infrastructures is also discussed.

Keywords Sensing networks • IoT • Smart Buildings

M. Arnesano (✉) • S. A. Mansi
Università Telematica eCampus, Novedrate, Italy
e-mail: marco.arnesano@uniecampus.it; silviaangela.mansi@uniecampus.it

© The Author(s) 2023
T. Lynn et al. (eds.), *Disrupting Buildings*, Palgrave Studies in
Digital Business & Enabling Technologies,
https://doi.org/10.1007/978-3-031-32309-6_2

2.1   Introduction

Sensors play a pivotal role in reducing buildings' energy demand and in reaching the near Zero Energy Building (nZEB) standard through deep renovation. From this perspective, buildings should become a system that "provides every occupant with productive, cost effective and environmentally approved conditions through continuous interaction among its elements" (Buckman et al., 2014, p. 96). The Internet of Things (IoT) represents the cornerstone in the definition of a smart building (SB). Using sensor networks, SBs provide the possibility for monitoring and managing energy consumption and indoor environmental quality (IEQ) (Minoli et al., 2017).

This chapter aims to (1) define the role of sensor networks in the field of deep renovation, (2) summarise the main advantages and benefits as well as (3) the main challenges and concerns associated with the use of sensor infrastructures.

The remainder of this chapter is structured as follows. Section 2.2 introduces the key concepts for the definition of construction sites, smart buildings and services based on sensor networks. Given the importance of IoT for the deployment of innovative sensing solutions in both construction sites and smart buildings; Sect. 2.2 also includes the description of the IoT architecture together with the main communication technologies. Then, Sect. 2.3 presents various sensing use cases for the construction and renovation stages and Sect. 2.4 presents the application of sensors in smart buildings. This is followed by a discussion of ethical and legal aspects related to the use of sensor data in Sect. 2.5 before concluding.

2.2   Key Definitions, Technologies and Approaches

2.2.1   The Role of Sensors on Construction Sites

Effective monitoring of construction sites allows managers to record progress at different stages and ensure that the project stays on schedule. With the advent of sensor technology and IoT, several activities in a construction site can be monitored automatically and in real-time. The interaction between multiple stakeholders can provide a better understanding of the status of different construction activities while improving productivity and saving time and cost. Through IoT, communication and positioning technologies can also improve safety (Zhao et al., 2021) and waste management (Sartipi, 2020).

3D point cloud data represents the main approach *to mapping and monitoring* construction progress in real-life large-scale projects. This measurement system can create a digital twin of the building and identify objects present on site, capturing their external surface. Improved knowledge can be obtained with *sensor integration*, a common procedure to integrate data from different sensors to improve the quality and the accuracy of information acquired by each sensor individually. Typically, the integration involves data fusion between *mapping* sensors and *positioning and communication* sensors (Moselhi et al., 2020). While positioning sensors measure the distance travelled by a body starting from its reference position, communication sensors allow the communication between different devices. For instance, communication sensors, when integrated with positioning and mapping sensors, can be used to enhance the outdoor tracking of resources (Domdouzis et al., 2007) in a supply chain management system (Rajendranath, 2011) and the construction safety management (Park et al., 2019). The main positioning technologies are presented in Table 2.1.

Different approaches can be applied for real-time monitoring construction progress. Table 2.2 summarises the main monitoring methodologies that are currently used in real-life construction sites for understanding a particular scene, positioning objects and tracking objects.

Table 2.1   Positioning sensors in construction

Approach	*Description*
Identification and tracking devices	Radio frequency identification (RFID) is a technology for identification and data communication with devices
Inertial measurements units (IMU)	Accelerometer, gyroscope and magnetometer are used to calculate the device position
Global navigation satellite systems (GNSS)	GNSS use a satellite-based navigation system with global coverage to localise objects in outdoor spaces
Ultrasonic sensors	These sensors use ultrasound to measure time of flight (ToF) and calculate the distance
Infrared (IR) sensors	IR sensing technology detects the light reflected in the infrared region of the electromagnetic spectrum

Table 2.2   Monitoring methodologies for the construction site

Methodology	Definition
Scene understanding	
Classification	A process to assign a label to the scene based on a supervised machine learning (SML) algorithm (Maturana & Scherer, 2015)
Object detection	A process to localise objects of a specific category of data (He et al., 2017)
Segmentation	A method to divide data into segments representing objects (Chen et al., 2018)
Positioning methods	
Proximity	A method to detect mobile targets by using location coordinates of nearby sensors with a priorly measured position (Deak et al., 2012)
Triangulation and trilateration	Methods to estimate the target position using triangles' properties (Deak et al., 2012)
Fingerprinting	Uses a database of signals measured at known locations to estimate an object's location by comparing the current signal with the stored signals
Dead reckoning	A method to estimate the local motion of a moving target relative to a previously known position
Visual positioning	Estimates the position and orientation of sensors by using imagery and point clouds data (Zhang & Singh, 2015)
Tracking methods	
Active tracking	Tracks moving objects using different position methods (Deak et al., 2012)
Passive tracking	Involves surveying instruments such as robotic stations, stationary cameras or lidar sensors (Giancola et al., 2019)

2.2.2   Smart Buildings and Smart Homes

A smart building is an intelligent structure that is "*[…] expected to address both intelligence and sustainability issues by utilising computer and intelligent technologies to achieve the optimal combinations of overall comfort level and energy consumption*" (Wang et al., 2012, p. 260). An SB adapts its operation and physical form to a particular event before the event happens while maintaining its energy efficiency and occupant satisfaction (Buckman et al., 2014). SBs require many stakeholders and a lot of interconnected-embedded devices, automated systems and wireless technologies to be capable of communicating with the internal and external environment. Sensors play a pivotal role for an SB because of the need of measuring several quantities, belonging to different domains, which are required for each service deployed in an SB. Just to mention a few, electricity and heat

Table 2.3   The main services and requirements of a smart building

Service	Details	Sensing requirements
Location-based services	Identify occupants or resources' location for improving serviceability	Tracking the position of different objects within the building
Energy efficiency	Optimise the building energy consumption	Communicate with smart grids and measurement of energy loads
Facility management	Maintenance operation and control of building facilities to reduce operations and maintenance time and cost	Fine-grained sensor network for detailed real-time and long-term monitoring
Occupant comfort	Optimise ambient conditions according to occupants' needs for improving health and productivity	Multi-domain and multi-physics measurement for comfort monitoring

meters are required for energy monitoring, and environmental sensors (temperature, relative humidity, light, CO_2) are required for occupants' comfort measurement and control. Table 2.3 summarises the main functions and sensing requirements.

A *Smart Home* is the SB declination in a residential context. It is an environment equipped with technologies that make occupants' lives more convenient while preserving energy efficiency. Smart appliance solutions can cover different aspects of occupants' daily life such as air conditioning, lighting, home security, data privacy, entertainment, surveillance, detection, and assisting living (Wilson et al., 2015).

2.2.3   IoT Architecture

IoT is one of the most influencing innovations in the field of communication (Atzori et al., 2010). Its application in the built environment gives the possibility to make everyday objects intelligent and connected by means of sensing, networking, and processing capabilities (Jia et al., 2019). IoT architectures are generally described and arranged in *Perception, Network and Application* layers. The *Perception Layer* is the physical layer equipped with sensors for sensing and data collection. It detects environmental parameters and identifies other intelligent sensors in the physical space to share information to the upper layers. The *Network Layer*, as the term suggests, is responsible for processing and transmitting the raw data network technologies. The highest level is the *Application Layer*, which

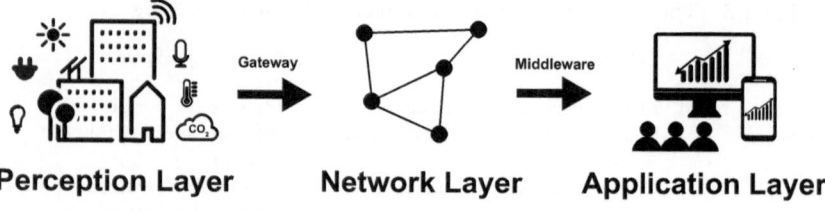

Fig. 2.1   Three-layers of an IoT architecture

creates the bridge between the building and the end user and supports the decision-making process. Figure 2.1 provides a schematic view of the three-layer IoT architecture.

Nevertheless, IoT needs to use a messaging and connectivity protocol to exchange information from remote locations. The recommended features of such protocol include (a) small code footprint (to be implemented in small devices), (b) low power consumption, (c) low bandwidth consumption, (d) low latency, and (e) use of a publish/subscribe (pub/sub) pattern. The most widespread messaging protocol is Message Queuing Telemetry Transport (MQTT), which is a lightweight pub/sub messaging transcription with a small footprint and minimal bandwidth (Spofford, 2019). Several communication protocols are available for the implementation of IoT architecture; Table 2.4 presents the most common ones.

2.3   Sensing During Construction and Renovation

Information handling is the most important aspect in industrial construction management (CM) (Wang et al., 2007). The main contexts of application in CM involve logistics, cost and time control, real-time process traceability, and operator safety (Ahmad et al., 2016). Recently, building information modelling (BIM) has been largely used in design, construction and facility management processes. The integration of sensors with BIM enables continuous monitoring of building construction stages for accurate construction and renovation management (e.g., cost and time). The integration of sensor data with BIM effectively creates a real-time digital twin that can continuously track changes and any discrepancies during the construction process. This enables the timely remediation of errors and monitoring of the condition of any material on-site by using cameras and other sensors (Liu et al., 2014). Sensor networks are also an

Table 2.4   IoT communication protocols

Concept	Definition
Short-range communication	
Bluetooth low energy (BLE)	A radio waves-based technology for communication between devices over short distances (2.402 GHz to 2.48 GHz)
Wi-Fi	A radio waves-based technology which allows the communication between smart devices according to IEEE 802.11 standards
UWB	A radio waves-based technology based on IEEE 802.15.4z-2020 standards, used for short-range (up to 200 metres) and high-bandwidth (500 MHz) communication
ZigBee	A high-level communication based on IEEE 802.15.4 specifications for creating personal area networks (PAN)
Long-range communication	
Frequency modulation (FM)	FM radio (88–108 MHz frequency) broadcast signals-based technology to localise the position by using fingerprinting techniques (Chen et al., 2013)
Long range (LoRa)	LoRa is an RF technology that uses a radio modulation technology for low-power, wide area network (LPWAN) communications
Sig Fox	A LPWAN communication technology, it uses Ultra-Narrow Bandwidth (UNB) modulation to send messages.
Cellular communication (CC)	CC is a way to connect people together for real-time communication and data transmission based on the global system for mobile communication standards (GSM)

efficient solution for supply chain *monitoring* and for managing building materials (Koskela & Vrijhoef, 1999). Table 2.5 the main sensor types that are currently used for mapping buildings for BIM creation and for monitoring construction sites.

BIM-sensor integration plays a key role in *preventive monitoring* during the construction phase to monitor and ensure proper structure conditions (Chen et al., 2020). IoT technologies can be used as a proactive tool to better predict building component failures, unplanned downtime, and broken tools, potentially increasing on-site productivity by up to 25% (Kayar et al., 2021). For automatic retrieval of physical information during the construction process, RFID has been widely adopted (Shen et al., 2010). BIM augmented with sensor data is also crucial for *facility management* (FM). The total cost of ownership of a building is heavily dependent on effective maintenance and the security and safety of the environment for the occupants. Several solutions have been proposed to monitor all building components to prevent failures and malfunctions

Table 2.5   Sensing construction: mapping sensors

Sensor type	Definition
Laser scanner (Lidar)	These sensors are based on the electromagnetic radiation near infrared spectrum for calculating the emitted pulse's time of flight (ToF) (El-Omari & Moselhi, 2011)
RGB camera	The structure from motion (SfM) approach is used for automated 3D reconstruction from digital images acquired by RGB cameras (Deak et al., 2012)
Depth camera	Cameras equipped with ToF sensors. One pulse of radiation illuminates the whole scene, thereby capturing all the reflected light. The range data are converted into an RBD-D image (Hübner et al., 2020)
Ground penetrating radar	Technologies used to map buried surface objects (Daniels, 2004)

(Cheng et al., 2020; Hemalatha et al., 2017). Finally, sensor networks can be used for monitoring the *end of life phase* at the end of the building life cycle. Tracking systems can be integrated into the building components pre-demolition to ensure traceability and the valorisation of waste (Dave et al., 2016).

2.4   SENSING DURING OPERATION: SMART BUILDINGS

Heating, ventilation and air conditioning (HVAC) systems account for about 40% of the energy consumption of a building and therefore their optimisation is critical for SB energy management. The main monitoring functions and the related applied sensors in an SB are presented in Table 2.6.

Recent developments in affordable *IEQ* sensors enable the continuous monitoring of indoor climate and to better analyse building performance. The IEQ monitoring approach consists of deploying many independent environmental sensors to measure air temperature, humidity, carbon dioxide (CO_2), particulate matter (PM), air pollutants, illuminance and noise (Choi et al., 2012). Serroni et al. (2021) developed a novel IoT system that includes an IR scanner and environmental sensors for monitoring pre- and post-renovation building performance. The monitoring of IEQ parameters, based on a non-intrusive IoT systems, allows the detection of building pathologies and such information can be used to support the renovation design. This can ultimately result in better performance

Table 2.6 Sensors applied to smart buildings

Monitoring function	Type of sensors
Energy	Electricity, heat meter, gas meters
IEQ	Temperature, humidity, CO_2, $PM_{2.5}$
Lighting system	IR sensor, lighting sensors
Security	IR sensor, video cameras, passive infrared (PIR)

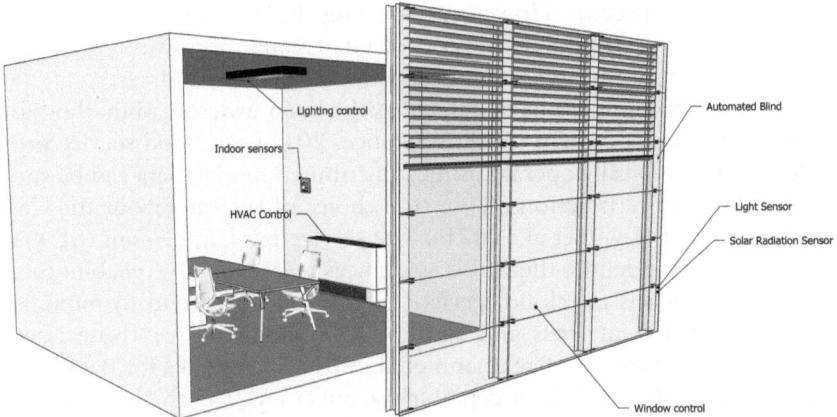

Fig. 2.2 Concept of the IoT façade module

post-renovation in terms of thermal comfort and indoor air quality. The application of IEQ monitoring to existing buildings can be facilitated with the integration of such sensors with plug and play façade modules for deep renovation. Arnesano et al. (2019), for example, propose the idea of a Smart-IoT façade. The panel is designed to embed sensors to measure indoor and outdoor conditions which are useful for the optimal control of the façade and HVAC (Fig. 2.2).

The advent of new wireless communication technologies and low-cost sensors is opening the possibility for accurate and fine-grained monitoring of the indoor environment in renovated buildings to provide HVAC and lighting systems with optimised control strategies. Kelly et al. (2013) integrated IoT and IEQ sensors in residential buildings implementing the communication between devices using the ZigBee protocol. Parkinson et al. (2019) developed a system consisting of low-cost sensors and a web platform for IEQ rating and analysis. These use cases provide evidence regarding the feasibility of advanced sensing implementation in existing buildings, thereby reducing cost and time for renovation.

2.5   CHALLENGES AND CONCERNS

The application of sensors in the deep renovation context is widely discussed in both research and industry as a way to increase the efficiency of the construction sector. However, managing the massive amount of data generated by both buildings and occupants creates a series of challenges and concerns that researchers and practitioners need to address.

Cloud computing has been defined as the main widespread method for sensors' data management (Mell & Grance, 2011). As cloud service providers (CSPs) typically operate using a distributed model, data can be subject to different jurisdictions. Thus, the choice of law can favour the CSP or the end user (Lynn et al., 2021). The service-level agreement (SLA) is the contract that defines the CSP's assurances on availability, reliability and performance levels for cloud service. In general, CSPs tend to minimise their liability for any loss and attempt to compensate for those issues through service credits (Bachmann et al., 2015). The rules for data management are defined by the acceptable use policy (AUP), which defines the prohibited activities and behaviours of the end users. It is important that AUP is aligned from both sides, CSP and clients, to avoid some issues (Hon et al., 2012).

Typically, CSPs are the "data processors" but also the "data controllers". The EU General Data Protection Regulation (GDPR) defines the data protection and privacy policies in case of accidental destruction, loss, or unauthorised disclosure of or access to personal data, without guaranteeing the integrity and availability of all data (Lynn et al., 2021). At the termination of the contract between a CPS and a client, an adequate provision for the subsequent handling of the client data needs to be provided (Bradshaw et al., 2011). In addition to the cloud-related data protection issues, challenges related to sensor data for automation in construction still represents a significant barrier mostly due to (1) a general lack of maturity in the use of information, (2) low level of investment in sensor technology and (3) difficulties in implementing effective communication and collaboration between stakeholders (Chen et al., 2018). In addition, industry is concerned about the lack of standardised practices for cost estimation and information security.

Considering the human side of the sensors, technologies for the end user should be capable of sensing environmental and personal contexts to ensure functional reliability. In smart homes where occupants need special assistance, failures or inaccurate inferences about the occupants' behaviour can lead to life-threatening consequences (Orpwood et al., 2005). There are significant concerns relating to privacy and security. For example, privacy could be compromised if data from IEQ sensors could provide information on the occupancy of the workplace (Cascone et al., 2017) or of the smart home (Cook, 2012). With regard to security, sensor data needs to be covered by legal protection, in case of malicious or unintentional data exposure (Sicari et al., 2015). Thus, when sensing technologies are developed, adequate consideration of the consequences of the data generated and their final use must be considered.

2.6   Conclusion

This chapter sheds light on sensing networks and their application for deep renovation, as well as considering the ethical and legal implications of their use. The introduction of sensing technologies presents opportunities to optimise and manage the construction and renovation process, from production to the end of life of buildings, and ideally both reduce costs and energy efficiency in existing buildings. Nevertheless, the huge amount of available data monitored represents a significant risk to data privacy and security. Those working with sensors and sensor data need to be knowledgeable about cybersecurity risks and appropriate mitigation measures.

References

Ahmad, M. W., Mourshed, M., Mundow, D., Sisinni, M., & Rezgui, Y. (2016). Building energy metering and environmental monitoring—A state-of-the-art review and directions for future research. *Energy and Buildings, 120*, 85–102. https://doi.org/10.1016/J.ENBUILD.2016.03.059

Arnesano, M., Bueno, B., Pracucci, A., Magnagni, S., Casadei, O., & Revel, G. M. (2019). Sensors and control solutions for Smart-IoT façade modules. *2019 IEEE International Symposium on Measurements and Networking, M and N 2019—Proceedings* (pp. 1–6). IEEE. https://doi.org/10.1109/IWMN. 2019.8805024

Atzori, L., Iera, A., & Morabito, G. (2010). The Internet of Things: A survey. *Computer Networks*, *54*(15), 2787–2805. https://doi.org/10.1016/J. COMNET.2010.05.010

Bachmann, R., Gillespie, N., & Priem, R. (2015). Repairing trust in organizations and institutions: Toward a conceptual framework. *Organization Studies*, *36*(9), 1123–1142. https://doi.org/10.1177/0170840615599334

Bradshaw, S., Millard, C., & Walden, I. (2011). Contracts for clouds: Comparison and analysis of the Terms and Conditions of cloud computing services. *International Journal of Law and Information Technology*, *19*(3), 187–223. https://doi.org/10.1093/IJLIT/EAR005

Buckman, A. H., Mayfield, M., & Beck, S. B. M. (2014). What is a smart building? *Smart and Sustainable Built Environment*, *3*(2), 92–109. https://doi. org/10.1108/SASBE-01-2014-0003

Cascone, Y., Ferrara, M., Giovannini, L., & Serale, G. (2017). Ethical issues of monitoring sensor networks for energy efficiency in smart buildings: A case study. *Energy Procedia*, *134*(October), 337–345. https://doi.org/10.1016/j. egypro.2017.09.540

Chen, J., Wu, J., & Qu, Y. (2020). Monitoring construction progress based on 4D BIM technology. *IOP Conference Series: Earth and Environmental Science*, *455*(1). https://doi.org/10.1088/1755-1315/455/1/012034

Chen, L. C., Papandreou, G., Kokkinos, I., Murphy, K., & Yuille, A. L. (2018). DeepLab: Semantic image segmentation with deep convolutional nets, atrous convolution, and fully connected CRFs. *IEEE Transactions on Pattern Analysis and Machine Intelligence*, *40*(4), 834–848. https://doi.org/10.1109/ TPAMI.2017.2699184

Chen, Y., Lymberopoulos, D., Liu, J., & Priyantha, B. (2013). Indoor localization using FM signals. *IEEE Transactions on Mobile Computing*, *12*(8), 1502–1517. https://doi.org/10.1109/TMC.2013.58

Cheng, J. C. P., Chen, W., Chen, K., & Wang, Q. (2020). Data-driven predictive maintenance planning framework for MEP components based on BIM and IoT using machine learning algorithms. *Automation in Construction*, *112*(January), 103087. https://doi.org/10.1016/j.autcon.2020.103087

Choi, J. H., Loftness, V., & Aziz, A. (2012). Post-occupancy evaluation of 20 office buildings as basis for future IEQ standards and guidelines. *Energy and Buildings*, *46*, 167–175. https://doi.org/10.1016/J.ENBUILD.2011.08.009

Cook, D. J. (2012). How smart is your home? *Science*, *335*(6076), 1579–1581. https://doi.org/10.1126/SCIENCE.1217640

Daniels, D. J. (2004). *Ground penetrating radar*. The Institution of Electronic Engineering and Technology.

Dave, B., Kubler, S., Främling, K., & Koskela, L. (2016). Opportunities for enhanced lean construction management using Internet of Things standards. *Automation in Construction*, *61*, 86–97. https://doi.org/10.1016/ J.AUTCON.2015.10.009

Deak, G., Curran, K., & Condell, J. (2012). A survey of active and passive indoor localisation systems. *Computer Communications, 35*(16), 1939–1954. https:// doi.org/10.1016/J.COMCOM.2012.06.004

Domdouzis, K., Kumar, B., & Anumba, C. (2007). Radio-Frequency Identification (RFID) applications: A brief introduction. *Advanced Engineering Informatics, 21*(4), 350–355. https://doi.org/10.1016/J.AEI.2006.09.001

El-Omari, S., & Moselhi, O. (2011). Integrating automated data acquisition technologies for progress reporting of construction projects. *Automation in Construction, 20*(6), 699–705. https://doi.org/10.1016/J.AUTCON. 2010.12.001

Giancola, S., Zarzar, J., & Ghanem, B. (2019). Leveraging shape completion for 3D Siamese tracking. In *Proceedings of the IEEE Computer Society Conference on Computer Vision and Pattern Recognition, 2019-June* (pp. 1359–1368). IEEE. https://doi.org/10.1109/CVPR.2019.00145

He, K., Gkioxari, G., Dollar, P., & Girshick, R. (2017). Mask R-CNN. In *Proceedings of the IEEE International Conference on Computer Vision, 2017-October* (pp. 2980–2988). IEEE. https://doi.org/10.1109/ICCV. 2017.322

Hemalatha, C., Rajkumar, M. V., & Gayathri, M. (2017). IoT based building monitoring system using GSM technique. *IOSR-Journal of Electronics and Communication Engineering (IOSR-JECE), 12*(2), 68–75.

Hon, W. K., Millard, C., & Walden, I. (2012). Negotiating cloud contracts—Looking at clouds from both sides now. *SSRN Electronic Journal.* https://doi. org/10.2139/SSRN.2055199

Hübner, P., Clintworth, K., Liu, Q., Weinmann, M., & Wursthorn, S. (2020). Evaluation of HoloLens tracking and depth sensing for indoor mapping applications. *Sensors, 20*(4), 1021. https://doi.org/10.3390/S20041021

Jia, M., Komeily, A., Wang, Y., & Srinivasan, R. S. (2019). Adopting Internet of Things for the development of smart buildings: A review of enabling technologies and applications. *Automation in Construction, 101*(January), 111–126. https://doi.org/10.1016/j.autcon.2019.01.023

Kayar, A., Öztürk, F., & Ceyhan, H. (2021). Increasing productivity and quality with IoT technologies in industrial treatment systems. In *Artificial intelligence systems and the Internet of Things in the digital era* (pp. 181–188). Springer. https://doi.org/10.1007/978-3-030-77246-8_18

Kelly, S. D. T., Suryadevara, N. K., & Mukhopadhyay, S. C. (2013). Towards the implementation of IoT for environmental condition monitoring in homes. *IEEE Sensors Journal, 13*(10), 3846–3853. https://doi.org/10.1109/JSEN. 2013.2263379

Koskela, L., & Vrijhoef, R. (1999). Roles of supply chain management in construction. In *Proceedings of the 7th annual conference of the International Group for Lean Construction* (pp. 133–146). IGLC.

Liu, Y.-F., Cho, S., Spencer, B. F., Jr., & Fan, J.-S. (2014). Concrete crack assessment using digital image processing and 3D scene reconstruction. *Journal of Computing in Civil Engineering, 30*(1). https://doi.org/10.1061/(ASCE) CP.1943-5487.0000446

Lynn, T., Mooney, J. G., van der Werff, L., & Fox, G. (2021). *Data privacy and trust in cloud computing: Building trust in the cloud through assurance and accountability.* Springer.

Maturana, D., & Scherer, S. (2015). VoxNet: A 3D Convolutional Neural Network for real-time object recognition. In *IEEE International Conference on Intelligent Robots and Systems, 2015-December* (pp. 922–928). IEEE. https://doi. org/10.1109/IROS.2015.7353481

Mell, P. M., & Grance, T. (2011). *The NIST definition of cloud computing.* https:// doi.org/10.6028/NIST.SP.800-145

Minoli, D., Sohraby, K., & Occhiogrosso, B. (2017). IoT considerations, requirements, and architectures for smart buildings-energy optimization and next-generation building management systems. *IEEE Internet of Things Journal, 4*(1), 269–283. https://doi.org/10.1109/JIOT.2017.2647881

Moselhi, O., Bardareh, H., & Zhu, Z. (2020). Automated data acquisition in construction with remote sensing technologies. *Applied Sciences, 10*(8), 2846. https://doi.org/10.3390/APP10082846

Orpwood, R., Gibbs, C., Adlam, T., Faulkner, R., & Meegahawatte, D. (2005). The design of smart homes for people with dementia—User-interface aspects. *Universal Access in the Information Society, 4*(2), 156–164. https://doi. org/10.1007/S10209-005-0120-7

Park, M., Park, S., Song, M., & Park, S. (2019). IoT-based safety recognition service for construction site. In *International Conference on Ubiquitous and Future Networks, ICUFN, 2019-July* (pp. 738–741). IEEE. https://doi. org/10.1109/ICUFN.2019.8806080

Parkinson, T., Parkinson, A., & de Dear, R. (2019). Continuous IEQ monitoring system: Context and development. *Building and Environment, 149,* 15–25. https://doi.org/10.1016/J.BUILDENV.2018.12.010

Rajendranath, A. S. (2011). *Performing need analysis and design of training on global navigation satellite system related applications for GRACE.* https://www. researchgate.net/publication/291836880_Performing_Need_Analysis_ and_Design_of_Training_on_Global_Navigation_Satellite_System_Related_ Applications_for_GRACE

Sartipi, F. (2020). Influence of 5G and IoT in construction and demolition waste recycling—Conceptual smart city design. *Journal of Construction Materials, 1*(4), 10.36756/JCM.V1.4.1.

Serroni, S., Arnesano, M., Violini, L., & Revel, G. M. (2021). An IoT measurement solution for continuous indoor environmental quality monitoring for buildings renovation. *Acta IMEKO, 10*(4), 230–238. https://doi.org/10.21014/ acta_imeko.v10i4.1182

Shen, W., Hao, Q., Mak, H., Neelamkavil, J., Xie, H., Dickinson, J., Thomas, R., Pardasani, A., & Xue, H. (2010). Systems integration and collaboration in architecture, engineering, construction, and facilities management: A review. *Advanced Engineering Informatics, 24*(2), 196–207. https://doi.org/10.1016/J.AEI.2009.09.001

Sicari, S., Rizzardi, A., Grieco, L. A., & Coen-Porisini, A. (2015). Security, privacy and trust in Internet of Things: The road ahead. *Computer Networks, 76,* 146–164. https://doi.org/10.1016/J.COMNET.2014.11.008

Spofford, D. (2019). *What is MQTT in IoT?* Retrieved June 6, 2022, from https://www.verypossible.com/insights/what-is-mqtt-in-iot

Wang, L. C., Lin, Y. C., & Lin, P. H. (2007). Dynamic mobile RFID-based supply chain control and management system in construction. *Advanced Engineering Informatics, 21*(4), 377–390. https://doi.org/10.1016/J.AEI.2006.09.003

Wang, Z., Wang, L., Dounis, A. I., & Yang, R. (2012). Integration of plug-in hybrid electric vehicles into energy and comfort management for smart building. *Energy and Buildings, 47,* 260–266.

Wilson, C., Hargreaves, T., & Hauxwell-Baldwin, R. (2015). Smart homes and their users: A systematic analysis and key challenges. *Personal and Ubiquitous Computing, 19*(2), 463–476. https://doi.org/10.1007/S00779-014-0813-0

Zhang, J., & Singh, S. (2015). *LOAM: Lidar odometry and mapping in real-time.* https://doi.org/10.15607/RSS.2014.X.007

Zhao, Z., Shen, L., Yang, C., Wu, W., Zhang, M., & Huang, G. Q. (2021). IoT and digital twin enabled smart tracking for safety management. *Computers and Operations Research, 128.* https://doi.org/10.1016/J.COR.2020.105183

CHAPTER 3

Building Information Modelling

Omar Doukari, Mohamad Kassem, and David Greenwood

Abstract From its origins as a computer-aided three-dimensional modelling tool, Building Information Modelling (BIM) has evolved to incorporate time scheduling, cost management, and ultimately an information management framework that has the potential to enhance decision-making throughout the whole life-cycle of built assets. This chapter summarises state-of-the-art BIM and its benefits. It then considers the particular characteristics of deep renovation projects, the challenges confronting their delivery, and the potential for using BIM to meet the challenges. This includes the application of Artificial Intelligence (AI) and Machine Learning (ML) to BIM models to optimise deep renovation project delivery. The prospects for this are encouraging, but further development work, including the creation of ontologies that are appropriate for renovation work, is still needed.

O. Doukari (✉) • D. Greenwood
Department of Mechanical and Construction Engineering, Northumbria University, Newcastle upon Tyne, UK
e-mail: omar.doukari@northumbria.ac.uk; david.greenwood@northumbria.ac.uk

M. Kassem
School of Engineering, Newcastle University,
Newcastle upon Tyne, UK
e-mail: mohamad.kassem@newcastle.ac.uk

Keywords BIM applications • 4D BIM • BIM dimensions • Deep renovation projects

3.1   INTRODUCTION

'BIM' can refer to an item (i.e., a building information model) as in its description by the US National Institute of Building Sciences as 'a digital representation of the physical and functional characteristics of a facility' (National Institute of Building Sciences, 2021). BIM can also be the process of managing construction information. This is defined by Hamil (2022) as '[...] creating and managing the information on a project throughout its whole life cycle'. Succar and Kassem (2015) have observed that BIM is a byword for digital innovation in construction. The concept of BIM first emerged in the early 1990s when earlier Computer-Aided Design (CAD) and 3D CAD software systems evolved into object-oriented 3D design tools containing geometric as well as non-graphical data. The term first became used in the early 2000s, notably in a white paper by Autodesk (2002). Nearly two decades after its first coinage as 'BIM', it has become a framework for managing information across the whole life cycle of projects as evidenced by the ISO 19650-1:2018 and ISO 19650-2:2018 standards and other related standards and guidance.

However, BIM has not permeated every part of the industry (Hamil & Bain, 2021) and there has been a temptation to 'cherry pick' convenient elements of the technology, leaving many wider aspects of BIM overlooked and their benefits unexploited (Georgiadou, 2019). There is also uncertainty over what BIM adoption actually means. Industry surveys predominantly reflect the use of BIM software, while academic studies tend to elicit the opinions of individual survey respondents. This has prompted attempts to measure BIM maturity. These range from the early Bew-Richards model comprising four levels of BIM (BSI, 2013) to more detailed multi-component approaches initiated by Succar (2009), further developed by Succar and Kassem (2015) to measure the maturity of countries or markets. Despite all the efforts, barriers to BIM adoption still remain. Begić and Galić (2021) found the most prominent of them to be resistance to change, required investment in software and skills, and cybersecurity concerns.

The remainder of this chapter is structured as follows. Section 3.2 explores BIM's benefits by examining its various applications through the project life cycle. The problems of delivering deep renovation projects are reviewed in Sect. 3.3, and then Sect. 3.4 considers how BIM can offer solutions. Section 3.5 considers some of the remaining challenges, and Sect. 3.6 offers a perspective on how current and future developments in BIM can overcome these challenges. Finally, Sect. 3.7 presents some concluding remarks.

3.2   BIM APPLICATIONS, BENEFITS, AND BEYOND

BIM originated as an enhanced 3D design tool but soon began to offer a wider range of functional applications that extended its range of use cases: so-called BIM dimensions. These offer the prospect of a unified model that can enable the efficient and effective sharing of data between different functions and throughout the project life cycle.

Table 3.1 presents a (non-exhaustive) list of the commonly recognised BIM dimension. In reality, once they go beyond 4D BIM, where a time schedule is linked to a 3D physical BIM model, these so-called dimensions

Table 3.1   Commonly recognised BIM 'dimensions'

Dimension	Description	Uses/benefits
3D BIM	A digital object-oriented representation with physical and functional information. Allows integration of multiple designs (architectural, structural, services, etc.)	Complements the distributed building design process. Enables parametric design, automatic code compliance checking, visualised renderings and 'walk-throughs', clash detection/resolution, and generation of off-site fabrication
4D BIM	BIM for scheduling and project delivery. Involves linking a time schedule to the 3D model to enhance and visualise construction planning techniques	Improves project management and communication between members of the project team through informative animations of the construction process
5D BIM	'5D BIM' offers automatic quantity take-off (QTO), cost management and analysis	Evaluation of the cost implications of design decisions and support for bidding, procurement, cost management, and accounting

Note: Further uses (safety, accessibility, security) have been proposed, with no consensus on terminology (Charef et al., 2018). The term 'nD model' reflects the range of possibilities (Aouad et al., 2006)

Table 3.2   Metaphoric 'dimensions' debated in literature

Dimension	Description	Uses/benefits
6D BIM	BIM for environmental sustainability. Incorporates information on embodied carbon, energy use, resource efficiency	Real-time feedback on the implications of design decisions, enabling detailed analysis of an asset's future performance
7D BIM	Focus on managing the operational life cycle of a built asset. The output to the owner or end user is in the form of an asset information model (AIM)	Can be linked to Building Management Systems (BMS) for functions such as predictive maintenance, facilities management, and building performance

are simply applications. As Koutamanis (2020) points out, time can realistically be considered a dimension, whereas cost (5D), sustainability (6D), or life cycle (7D) are metaphors. The terminology is nevertheless retained here as it is still widely recognised. Stepping aside from the BIM dimensions, a more inclusive coverage of the applications and uses of BIM is represented by the PennState BIM uses (2023) or the concept of model uses of Succar et al. (2016) (Table 3.2).

Related literature suggests that BIM can generate a number of organisational benefits. According to Georgiadou (2019), they include design optimisation, improved on-time delivery, cost efficiency, quality assurance, collaboration and communication, and sustainability. Ghaffarianhoseini et al. (2017) also add technical superiority, interoperability, information capture, improved cost control, whole-life applicability, the potential for integrated procurement, and reduced conflict and better communication and coordination within the project delivery team. Attempts to quantify the value of such benefits using, for example, a return on investment (ROI) approach are necessarily context-specific. This is confirmed by Sompolgrunk et al. (2021) that found a huge variance in reported ROIs. Positive results were predominantly associated with schedule reduction/compliance, increased productivity, and reduction in requests for information, change orders, and rework.

As highlighted in Begić and Galić (2021), BIM is a vital element in the transformation to 'Construction 4.0', where innovations such as the Internet of Things (IoT), blockchain, and artificial intelligence (AI) and modern methods of construction (MMC) will play an increasing role in the built environment, and built assets will have a golden thread of information showing how they have been built and how they are performing (Hamil, 2022).

The digital and object-oriented basis of BIM allows it to interact with other digitally driven systems which can represent either inputs to a BIM model (e.g., the retrospective modelling of existing facilities through point-cloud surveys) or outputs (e.g., the automated manufacture of building components from their design). Other examples include the integration of BIM with blockchain to overcome challenges related to provenance, accuracy, transparency, security, and ownership of model information (Li et al., 2019). Furthermore, an opportunity for transforming the management of built assets comes with the concept of the 'digital twin'—a cyber-physical system where live data flows from sensors[1] in the physical asset (e.g., a building) into its counterpart digital model (De Luca et al., 2021). Conversely, the physical twin can be controlled from the model to enable operation, maintenance, monitoring, diagnostics, prediction, and simulation. These activities can focus on such important issues as energy use, carbon emissions, and planned maintenance.

3.3   DEEP RENOVATION PROJECTS: KEY CHALLENGES

The delivery of construction projects in general can be complex and demanding and presents well-documented challenges to the control of cost, schedule, and quality. This situation becomes even more acute in the case of renovation projects, which are inherently more uncertain.

Planning and execution of deep renovation[2] projects are currently driven by judgement and experience rather than standardised solutions (Amorocho & Hartmann, 2021; Lynn et al., 2021). Such projects typically disturb existing building occupants, whose presence, conversely, disrupts construction logistics, schedules, and budgets. Deep renovation projects, which aim at maximising energy efficiency in the renovation process (Shnapp et al., 2013) are even more problematic because of their extended impact on the fabric, services, and even structure of a building (Fawcett, 2014). McKim et al. (2000) reported that renovation projects were twice as susceptible to delay and suffered four times the cost overruns of new construction work. Their conclusion was that conventional

---

[1] Chapter 2 in this book provides more details on the use of sensor networks in the context of deep renovation projects.

[2] Chapter 1 in this book provides a detailed definition of Deep Renovation.

time and cost control techniques were inadequate for such projects. Alongside the uncertainties surrounding the work itself is the safety and well-being of building occupants. Chaves et al. (2016) have highlighted disruptions involving: (a) utilities (gas, electricity, telecoms); (b) access; (c) use of space by both occupants and contractors; (d) problems with internal environmental quality (noise, dust, vibration, and debris); (e) external environmental quality; and (f) transport and parking spaces. To mitigate such issues and allow project teams to plan, organise, and efficiently realise renovation tasks, Killip et al. (2013) have suggested the adoption of new technologies and optimised processes: approaches that are epitomised by BIM-based applications. BIM benefits are much reported in literature but rarely in relation to renovation projects.

3.4   The Potential for BIM in Deep Renovation Projects

In their state-of-the-art review of design decision-making for sustainable renovation projects, Passoni et al. (2021) stress the need for multi-criteria decision methods, optioneering, and pre-validation of proposals. Although their work relates to the *design* of sustainable renovation projects, the conclusions apply equally to their delivery. In both cases digital tools based on BIM can be employed to identify, optimise, validate, and communicate different renovation scenarios, in terms of cost, time, and effectiveness in meeting the required functionality and quality of the resulting work.

A foundation for using BIM in most renovation projects is employing a laser scanner to capture point-cloud data that can then be processed to create a 3D BIM model (Wang & Kim, 2019). The retro-constructed geometric 3D BIM model can then be semantically enriched to enable further functionalities. Thus the 'scan to BIM' or 'mapping' stage can provide the basis for the application of BIM in renovation projects as described by D'Oca et al. (2018) in their review of related European Horizon 2020 projects. The captured as-built BIM models can be used for building condition assessment that can underpin the prioritisation of renovation options. For example, Sebastian et al. (2018) describe the use, based on the initial scanned model, of software applications for assessing conditions and analysing and prioritising renovation interventions on the basis of their energy performance. Acampa et al. (2021) have also shown how BIM-based decision support systems have been used to generate such optimal renovation scenarios.

BIM enables a common data environment (CDE) for information exchange between the various design consultants, connecting energy simulation and prediction (Garwood et al., 2018; Pinheiro et al., 2018), life cycle costing (Edwards et al., 2019; Sharif & Hammad, 2019) and its parametric modelling capacity enables quicker and more cost-effective design optimisations (Abanda & Byers, 2016; Corgnati et al., 2017).

From a project delivery perspective, BIM can enhance the management of project schedules through 4D BIM (Jupp, 2017; Sheikhkhoshkar et al., 2019) and budgetary control using '5D BIM' (Lee et al., 2016). The benefits of doing so include more critical assessment of options, more effective coordination and work sequencing, improved tracking and review, enhanced utilisation of space and resources, control of waste, and improved communication (Gledson & Greenwood, 2017). In deep renovation projects, these benefits are amplified. In fact, the uncertainty inherent in renovation work requires flexible approaches, and in this respect, the ability of BIM-based simulations of time and cost to evaluate different renovation scenarios offers great potential (Chaves et al., 2016).

Finally, from a communication perspective, the use of BIM simulation and visualisation can be useful in mitigating disruption to (and by) occupants (Passoni et al., 2021). Crucially, BIM simulations of time and cost can enable users to share and clarify the perception of the renovation process with all stakeholders, including building occupants who are unlikely to fully understand traditional drawings and schedules. The ability to use visualisation to demonstrate design and construction decisions and their consequences in time and space, including any different options that are available, can clearly facilitate good relations and better cooperation with owners and occupiers. This, in turn, should assist the renovation process.

3.5   IMMEDIATE CHALLENGES IN THE ADOPTION OF BIM SOLUTIONS

The prospects for the use of BIM in deep renovation projects are encouraging, but there remain challenges, some of which are related to interoperability and workflow. As observed by De Gaetani et al. (2020), the multidisciplinary nature of construction design attracts the use of different types of authoring software, file formats, or (even if formats are the same) different file format versions. Thus, additional effort may be required for file exchange. The development of time- and cost-related models from initial 3D design models is typically performed later in the project process by the contractor. Here the inherent advantage of BIM is the opportunity to extract objects from the design model to generate scheduled activities and budget items. As identified by Park and Cai (2015), this process involves four phases:

1. Extracting object information from the 3D model
2. Defining the appropriate work (WBS) or cost (CBS) breakdown structures for the project
3. Linking the elements of these with the objects in the BIM model
4. Generating the schedule or cost models themselves

In theory, this process could be automated (ElMenshawy & Marzouk, 2021), but this depends on the interoperability between the applications used and the degree of coordination with the original design. Design-oriented 3D BIM models are rarely set up to facilitate the subsequent production of schedule and budgetary tools. As a result, significant manual effort, involving the splitting or aggregating of objects, is required to link the elements of the 3D model to the relevant time and cost parameters. Such interoperability and workflow challenges must be overcome to unlock the full efficiencies of information transfer enabled by BIM adoption.

3.6   Further Developments and Challenges

Furthermore, the availability of BIM models for delivering renovation projects presents an opportunity to exploit advances in Artificial Intelligence (AI) and Machine Learning (ML). Mulero-Palencia et al. (2021) raise the possibility of applying AI and ML to BIM models of deep renovation interventions. Their focus is on the development of algorithms for diagnosis (at the preliminary analysis stage of a project) and optimisation (at the design simulation and optioneering stage), but there are implications for all stages in the life cycle of deep renovation projects.

A current barrier to doing so is the lack of ontologies that are appropriate for renovation work. Ontologies are fundamental requirements for formalising specific domain knowledge including concepts, relations, and constraints and are thus an essential basis for producing machine-readable code that can support process automation (Hartmann & Trappey, 2020). As noted by Amorocho and Hartmann (2021), BIM-based tools for design, planning, and project management are normally targeted at new construction and comprehensive ontologies for renovation activities are not currently available. Amorocho and Hartmann (2021) have developed a limited example ontology that was restricted to the installation of common renovation products, such as windows, HVAC components, and external thermal insulation panels. However, no ontology currently exists for the general case of renovation projects, and further development will be necessary to capture the potential of BIM-driven AI solutions.

3.7   CONCLUSION

This chapter summarises the state-of-the-art on BIM with a specific emphasis on its potential in deep renovation projects. In the course of domestic renovation works, disruptions to users and occupants are inevitable. This and the uncertainties inherent in this type of work make the delivery of such projects challenging—particularly in adhering to time and cost plans. Deep renovations are especially problematic in this respect. The use of BIM would not only enable the integration of condition assessments with subsequent building design, but also permit the automatic extraction of design information to generate schedules and budgets and to control them. Based on BIM, other technologies, such as AI and ML, could be applied to generate standardised and optimised solutions for the delivery of deep building renovation projects.

REFERENCES

Abanda, F. H., & Byers, L. (2016). An investigation of the impact of building orientation on energy consumption in a domestic building using emerging BIM (Building Information Modelling). *Energy, 97,* 517–527.

Acampa, G., Diana, L., Marino, G., & Marmo, R. (2021). Assessing the transformability of public housing through BIM. *Sustainability, 13*(10), 5431.

Amorocho, J. A. P., & Hartmann, T. (2021). Reno-Inst: An ontology to support renovation projects planning and renovation products installation. *Advanced Engineering Informatics, 50,* 101415.

Aouad, G., Lee, A., & Wu, S. (2006). *Constructing the future: nD modelling.* Routledge.

Autodesk. (2002). *Building Information Modeling.* Autodesk Building Industry Solutions. Retrieved April 18, 2022, from http://www.laiserin.com/features/bim/autodesk_bim.pdf

Begić, H., & Galić, M. (2021). A systematic review of Construction 4.0 in the context of the BIM 4.0 premise. *Buildings, 11*(8), 337.

BSI. (2013). *PAS 1192-2: Specification for information management for the capital/delivery phase of construction projects using Building Information Modelling*. BSI Standards Limited.

Charef, R., Alaka, H., & Emmitt, S. (2018). Beyond the third dimension of BIM: A systematic review of literature and assessment of professional views. *Journal of Building Engineering, 19*, 242–257.

Chaves, F. J., Tzortzopoulos, P., Formoso, C. T., & Biotto, C. N. (2016). Building Information Modelling to cut disruption in housing retrofit. *Proceedings of the Institution of Civil Engineers-Engineering Sustainability, 170*(6), 322–333.

Corgnati, S. P., Cotana, F., D'Oca, S., Pisello, A. L., & Rosso, F. (2017). A cost-effective human-based energy-retrofitting approach. In *Cost-effective energy efficient building retrofitting* (pp. 219–255). Woodhead Publishing.

D'Oca, S., Ferrante, A., Ferrer, C., Pernetti, R., Gralka, A., Sebastian, R., & op 't Veld, P. (2018). Technical, financial, and social barriers and challenges in deep building renovation: Integration of lessons learned from the H2020 cluster projects. *Buildings, 8*(12), 174.

De Gaetani, C. I., Mert, M., & Migliaccio, F. (2020). Interoperability analyses of BIM platforms for construction management. *Applied Sciences, 10*(13), 4437.

De Luca, D., Dettori, M., Del Giudice, M., & Osello, A. (2021). Connected BIM models towards Industry 4.0. In *Handbook of research on developing smart cities based on digital twins* (pp. 219–242). IGI Global.

Edwards, R. E., Lou, E., Bataw, A., Kamaruzzaman, S. N., & Johnson, C. (2019). Sustainability-led design: Feasibility of incorporating whole-life cycle energy assessment into BIM for refurbishment projects. *Journal of Building Engineering, 24*, 100697.

ElMenshawy, M., & Marzouk, M. (2021). *Automated BIM schedule generation approach for solving time–cost trade-off problems*. Engineering, Construction and Architectural Management.

Fawcett, T. (2014). Exploring the time dimension of low carbon retrofit: Owner-occupied housing. *Building Research and Information, 42*(4), 477–488.

Garwood, T. L., Hughes, B. R., Oates, M. R., O'Connor, D., & Hughes, R. (2018). A review of energy simulation tools for the manufacturing sector. *Renewable and Sustainable Energy Reviews, 81*, 895–911.

Georgiadou, M. C. (2019). *An overview of benefits and challenges of Building Information Modelling (BIM) adoption in UK residential projects*. Construction Innovation.

Ghaffarianhoseini, A., Tookey, J., Ghaffarianhoseini, A., Naismith, N., Azhar, S., Efimova, O., & Raahemifar, K. (2017). Building Information Modelling (BIM) uptake: Clear benefits, understanding its implementation, risks and challenges. *Renewable and Sustainable Energy Reviews, 75*, 1046–1053.

Gledson, B. J., & Greenwood, D. (2017). *The adoption of 4D BIM in the UK construction industry: An innovation diffusion approach.* Engineering, Construction and Architectural Management.

Hamil, S. (2022). *What is BIM?* NBS Enterprises. Retrieved April 30, 2022, from https://www.thenbs.com/knowledge/what-is-building-information-modelling-bim

Hamil, S., & Bain, D. (2021). *Digital construction report 2021.* NBS Enterprises. Retrieved June 9, 2022, from https://www.thenbs.com/digital-construction-report-2021/

Hartmann, T., & Trappey, A. (2020). Advanced Engineering Informatics - Philosophical and methodological foundations with examples from civil and construction engineering. *Developments in the Built Environment,* 4100020. https://doi.org/10.1016/j.dibe.2020.100020

Jupp, J. (2017). 4D BIM for environmental planning and management. *Procedia Engineering, 180,* 190–201.

Killip, G., Janda, K., & Fawcett, T. (2013). *Building Expertise: industry responses to the low-energy housing retrofit agenda in the UK and France,* presented at the Conference ECEEE Summer Study, Presqu'ile de Giens, France.

Koutamanis, A. (2020). Dimensionality in BIM: Why BIM cannot have more than four dimensions? *Automation in Construction,* 114103153. https://doi.org/10.1016/j.autcon.2020.103153

Lee, X. S., Tsong, C. W., & Khamidi, M. F. (2016). 5D Building Information Modelling—A practicability review. In *MATEC web of conferences* (vol. 66, p. 26). EDP Sciences.

Li, J., Greenwood, D., & Kassem, M. (2019). Blockchain in the built environment and construction industry: A systematic review, conceptual models and practical use cases. *Automation in Construction, 102,* 288–307.

Lynn, T., Rosati, P., Egli, A., Krinidis, S., Angelakoglou, K., Sougkakis, V., Tzovaras, D., Kassem, M., Greenwood, D., & Doukari, O. (2021). RINNO: Towards an open renovation platform for integrated design and delivery of deep renovation projects. *Sustainability, 13*(11), 6018.

McKim, R., Hegazy, T., & Attalla, M. (2000). Project performance control in reconstruction projects. *Journal of Construction Engineering and Management, 126*(2), 137–141.

Mulero-Palencia, S., Álvarez-Díaz, S., & Andrés-Chicote, M. (2021). Machine learning for the improvement of deep renovation building projects using as-built BIM models. *Sustainability, 13*(12), 6576.

National Institute of Building Sciences. (2021). National BIM Standard-United States- V3. Retrieved April 30, 2022, from https://www.nationalbimstandard.org/

Park, J., & Cai, H. (2015). Automatic construction schedule generation method through BIM model creation. In *International workshop on computing in civil engineering* (pp. 620–627). ASCE Library.

Passoni, C., Marini, A., Belleri, A., & Menna, C. (2021). Redefining the concept of sustainable renovation of buildings: State of the art and an LCT-based design framework. *Sustainable Cities and Society, 64*, 102519.

PennState. (2023). *BIM Uses*. PennState College of Engineering. Retrieved April 18, 2022, from https://bim.psu.edu/uses/

Pinheiro, S., Wimmer, R., O'Donnell, J., Muhic, S., Bazjanac, V., Maile, T., et al. (2018). MVD based information exchange between BIM and building energy performance simulation. *Automation in Construction, 90*, 91–103.

Sebastian, R., Gralka, A., Olivadese, R., Arnesano, M., Revel, G. M., Hartmann, T., & Gutsche, C. (2018). Plug-and-play solutions for energy-efficiency deep renovation of European building stock. *Multidisciplinary Digital Publishing Institute Proceedings, 2*(15), 1157.

Sharif, S. A., & Hammad, A. (2019). Simulation-based multi-objective optimization of institutional building renovation considering energy consumption, life-cycle cost and life-cycle assessment. *Journal of Building Engineering, 21*, 429–445.

Sheikhkhoshkar, M., Rahimian, F. P., Kaveh, M. H., Hosseini, M. R., & Edwards, D. J. (2019). Automated planning of concrete joint layouts with 4D-BIM. *Automation in Construction, 107*, 102943.

Shnapp, S., Sitjà, R., & Laustsen, J. (2013). *Deep renovation definition*. Global Buildings Performance Network, Technical Report, February 2013. Retrieved April 19, 2022, from https://www.gbpn.org/wp-content/uploads/2021/06/08.DR_TechRep.low_.pdf

Sompolgrunk, A., Banihashemi, S., & Mohandes, S. R. (2021). Building information modelling (BIM) and the return on investment. *A systematic analysis Construction Innovation, 23*(1), 129–154. https://doi.org/10.1108/CI-06-2021-0119

Succar, B. (2009). Building information modelling framework: A research and delivery foundation for industry stakeholders. *Automation in Construction, 18*(3), 357–375.

Succar, B., & Kassem, M. (2015). Macro-BIM adoption: Conceptual structures. *Automation in Construction, 57*, 64–79.

Succar, B., Saleeb, N., Sher, W. (2016, July). Model Uses: Foundations for a Modular Requirements Clarification Language. In *Proceedings of the Australasian Universities Building Education (AUBEA2016) Conference*. Cairns, Australia.

Wang, Q., & Kim, M. K. (2019). Applications of 3D point cloud data in the construction industry: A fifteen-year review from 2004 to 2018. *Advanced Engineering Informatics, 39*, 306–319.

CHAPTER 4

Building Performance Simulation

*Asimina Dimara, Stelios Krinidis, Dimosthenis Ioannidis,
and Dimitrios Tzovaras*

Abstract Simulation is a proven technique that uses computational, mathematical, and machine learning models to represent the physical characteristics, expected or actual operation, and control strategies of a building and its energy systems. Simulations can be used in a number of tasks along the deep renovation life cycle, including: (a) integrating simulations with other knowledge-based systems to support decision-making, (b) using simulations to evaluate and compare design scenarios, (c) integrating simulations with real-time monitoring and diagnostic systems for building energy management and control, (d) integrating multiple

A. Dimara • D. Ioannidis • D. Tzovaras
Information Technologies Institute, Centre of Research & Technology Hellas (CERTH), Thessaloniki, Greece
e-mail: adimara@iti.gr; djoannid@iti.gr; Dimitrios.Tzovaras@iti.gr

S. Krinidis (✉)
Information Technologies Institute, Centre of Research & Technology Hellas (CERTH), Thessaloniki, Greece

Department of Management Science and Technology, International Hellenic University, Kavala, Greece
e-mail: krinidis@iti.gr

© The Author(s) 2023   53
T. Lynn et al. (eds.), *Disrupting Buildings*, Palgrave Studies in
Digital Business & Enabling Technologies,
https://doi.org/10.1007/978-3-031-32309-6_4

simulation applications, and (e) using virtual reality (VR) to enable digital building design and operation experiences. While building performance simulation is relatively well established, there are numerous challenges to applying it across the renovation life cycle, including data integration from fragmented building systems, and modelling human-building interactions, amongst others. This chapter defines the building performance simulation domain outlining significant use cases, widely used simulation tools, and the challenges for implementation.

Keywords Building simulation • Building performance simulation • Building simulation applications

4.1   INTRODUCTION

Building simulation (BS) is the process of creating a digital replica of a building, while building performance simulation (BPS) is a model that evaluates how the building performs under real-life conditions (Mahdavi, 2020). During the replication process, digital copies are created of the whole building—its exterior and interior, and, in some cases, the building's distinct parts (e.g., apartments and rooms) if needed. The BPS process consists of five main phases as depicted in Fig. 4.1. The main objective of this process is to define the best performance criteria and the most suitable actions by applying performance simulations. Once results are generated, they are evaluated against initial expectations and requirements.

Mathematical and intelligent models and applications are exploited to recreate (simulate) various external and internal conditions while representing them in a virtual environment (Mahdavi, 2020). BPS makes it easier for different stakeholders (building managers, architects, engineers, etc.) to inspect and check salient points, elements, and other aspects of the building's life cycle (i.e., early design, construction, retrofitting, monitoring, inspection, and demolition) (Bramstoft et al., 2018).

Exploiting BS tools and applications is faster, safer, and less expensive than producing a real use case scenario. It supports product and system testing without having to build them in real life and is often less time-consuming and costly while also being safer. Moreover, BPS may be exploited for identifying building problems by replicating and producing different conditions. Finally, it may be used to model specific changes to check how the building reacts in the short or the long term (Fernandez-Antolin et al., 2022).

The remainder of the chapter is structured as follows: Sect. 4.2 describes the main applications of and approaches to BPS. Section 4.3

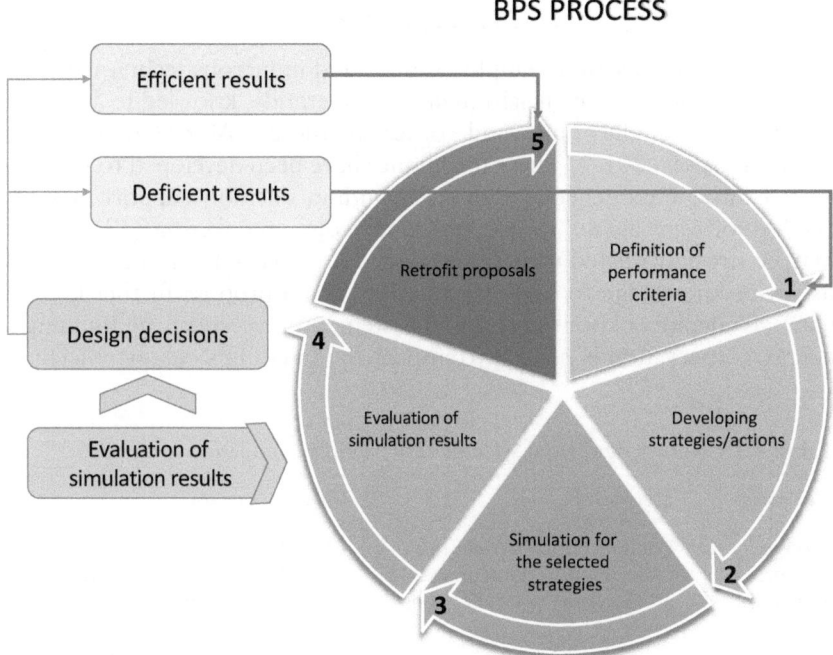

Fig. 4.1 Building performance simulation process (Bramstoft et al., 2018)

introduces BPS in more detail. Finally, Sect. 4.4 presents some concluding remarks.

4.2 Building Performance Simulation Approaches and Applications

BPS is a dynamic technique that is used to predict the behaviour of a building while optimising energy efficiency (Attia et al., 2013). The key objective of BPS is to reduce the building's environmental footprint while improving indoor environmental quality (IEQ). At the same time, if applied correctly, BPS may facilitate technological innovation and progress in building construction (Loonen et al., 2017). BPS energy models are applied in a number of real-life applications. These include, for example, energy conservation, energy monitoring, energy savings, and fault detection. Simulations may include load and energy simulation, energy management simulation, virtual reality simulations, and a wide range of other simulations based on stakeholders' needs (Martins, 2022).

4.2.1   Integrating Simulations with Other Knowledge-Based Decision Support Systems

Knowledge-based systems employ various and numerous techniques such as statistical analysis, artificial intelligence methods, knowledge and data visualisation, engineering, and other methods (Alor-Hernández & Valencia-García, 2017). These techniques have been developed to be integrated into heterogeneous systems including decision support systems (DSS), software agents (SAs), and knowledge engineering (KE) which may be strongly related to BS as described in Table 4.1. All these systems use prior knowledge to exclude, determine, and propose further knowledge. To deploy a knowledge-based system, an analysis of the building's energy conservation is needed to inspect the building's energy efficiency.

Table 4.1   Knowledge-based systems and exemplar simulation use cases

Technology	Description	Use cases examples
Decision support systems	Data-driven applications that facilitate the decision process by providing the optimal options	• Simulate the development of a building before it even starts to evaluate and support various alternatives • Simulations during retrofit using different building materials to support testing and evaluation • Simulation of thermal conditions and building behaviour to manage real-life observations
Software agents	Object-oriented application that performs specific functions to provide instinct on specific processes	• Simulation of dynamic building ecosystems to present new resource-optimisation approaches • Simulation of smart grids to showcase the behaviour of the grid to act independently • Simulation of the indoor environment to define optimal indoor conditions
Knowledge engineering	Machine learning applications that try to emulate human decisions based on expertise	• Simulate the building process sequence to determine the optimal sequence of specific actions • Simulate building modifications to compute a specific parameter • Simulation of building conditions to identify specific constraints

In general, knowledge-based systems are the main component of building simulation and are mainly used as the process that emulates the simulation outcome.

4.2.2   Using Simulations to Evaluate and Compare Design Scenarios

Simulations provide many advantages compared to a conceptual design. For example, BPS provides the ability to examine various solutions during the design stage like the efficiency of the building's equipment and integration, therefore reducing development time and energy consumption and emissions. Moreover, BPS may facilitate the optimisation of thermal and visual comfort by simulating the building's fenestration and massing. This process may also be deployed in the early stages while designing the façade. Finally, BPS enables the simulation of heating, ventilation, and air conditioning (HVAC) systems to define the optimal setup.

4.2.3   Integrating Simulations with Real-Time Monitoring and Diagnostic Systems for Building Energy Management and Control

Radiation, conduction, and convection are mass and heat transfer phenomena that take place in a building and are the key inputs for energy and load simulations (Yu, 2019). As a result, every BS needs to take into account different mechanisms behind such phenomena. Heat and mass transfer are carried in the building by air movements and are embodied by the indoor air pressure stack between outdoor and indoor places (Vera, 2018). Those phenomena are influenced by peoples' activities, heating and cooling systems, and ventilation as well as building insulation and orientation (Puttur et al., 2022). Therefore, thermal and airflow models are applied to represent heat and mass transfers of buildings as they are significantly associated with the energy transfer (Tian, 2018b) and are used to calculate loads and simulate energy consumption (Tan et al., 2022).

Frequently used models for simulating energy consumption include Computational Fluid Dynamics (CFD) models, zonal models, and multi-zone models (Laghmich et al., 2022). CFD models separate the building into cells to simulate load and energy consumption (Shen et al., 2020). A multi-zone model uses rooms as computational elements for the simulation, while a zonal model uses several zones by separating a room into

Table **4.2** Airflow models for energy simulation comparison (Laghmich et al., 2022)

Airflow models	Model complexity	Computational time and resources required	Level of accuracy reached	User experience/ knowledge needed	Size of minimum zone element
Multi-zone model	Low	Low	Low	Low	Large
Zonal mode mode	Medium	Medium	Medium	Medium	Medium
CFD model	High	High	High	High	Small

smaller units (Yu, 2019). An overall assessment of the aforementioned models is presented in Table 4.2.

Other models estimate a building's energy consumption by exploiting physical models (Kampelis et al., 2020). These models simulate energy consumption by exploiting mathematical equations and the building's energy conservation. While such mathematical or computational models are sufficiently precise, they require holistic building information (Oucquier et al., 2013). Furthermore, a unique model is required for each building. Another widely referenced approach to simulating building energy load is data-driven modelling (Bermeo-Ayerbe et al., 2022). These models use indoor monitoring and measurements (e.g., relative humidity, temperature, historical consumption data, historical load, and generations data) to predict energy consumption (Kampelis et al., 2020). A benchmarking analysis of various regression models using energy consumption suggests that, in many cases, these models are sufficiently accurate for building simulation and may be used as a generic solution (Dimara et al., 2021).

To manage the overall energy consumption of a building, various load controls are applied to manage both energy savings and comfort regulation. The energy load of all appliances in a building are simulated in order to build predictive models for energy consumption and deploy an accurate energy management strategy (Fanti et al., 2018). The main problem when trying to find optimal control states is to detect the best strategies for heating, cooling, ventilation, and lighting that result in energy savings while maintaining desirable indoor conditions for the occupants. As such, all possible energy load actions must be simulated accurately.

Building energy modelling and simulation allows stakeholders to better understand certain energy operating characteristics before designing, applying, or testing them in a real-life scenario. Furthermore, it helps with reducing waste and allows energy-saving verification by testing real data against various scenarios which may take into account multiple factors such as weather conditions and occupancy patterns.

4.2.4   *Integrating Multiple Simulation Applications*

As mentioned previously, most simulations require the deployment of multiple models, applications, and techniques to provide an overall building assessment. In general, deploying and integrating automated multi-simulation applications may produce significant advantages when compared to a single simulation. During this procedure, a couple of models are integrated and their output is combined. For example, a comprehensive evaluation of the HVAC systems in a building would require the combination of airflow, heat losses, atmospheric conditions, and energy performance simulations.

4.3   BUILDING PERFORMANCE SIMULATION USE CASES

Deployment of modelling and simulation tools for building performance can be implemented in various and numerous use cases from the design stage to operation and management of a building. Some of the most common BPS simulation use cases are summarised in Table 4.3.

To implement any of the use cases above or any type of building simulation, appropriate tools and technologies must be applied. Some indicative technologies and commercial tools for BPS are summarised in Tables 4.4 and 4.5.

4.4   BUILDING SIMULATION: CHALLENGES AND CONCERNS

In the summary of literature on BPS, Attia (2010) identifies five major challenges—(a) interface usability and information management, (b) integration of decision design support and design optimisation, (c) accuracy and ability to simulate detailed and complex building components, (d) integration with other tools in the building design and construction/renovation process, and (e) BIM[1] integration and interoperability. These

---

[1] Chapter 3 in this book provides a detailed discussion on BIM.

Table 4.3   Simulation use cases along the deep renovation life cycle

Use case	Construction/deep renovation life cycle
Simulation of the building performance will facilitate the decision-making for designing and installing specific building components. Specifically, simulation of the optimal performance for the user's comfort and energy consumption	• Design process • Retrofit • Operation and monitoring
Baseline for certain energy certification schemes. Compare the building's energy consumption baseline to energy-saving to obtain the correct designs for green energy certifications like BREEAM (Liu et al., 2021)	• Design process • Retrofit • Operation and monitoring
Evaluation of various systems, applications, and systems. Creating a bidding virtual prototype to assess different materials, applications, and systems to identify all the possibilities for future refinement	• Retrofit • Renovation • End of cycle
Visualisation of control strategies. BPS to visualise promising energy control strategies to achieve the optimal building performance (GhaffarianHoseini et al., 2017)	• Operation and monitoring
Fine-tuning of HVAC. Simulation of the HVAC system energy performance and coefficients are of key importance during the decision of selecting the HVAC system and installation (Kamel & Memari, 2019)	• Design • Operation and monitoring
Visualisation of optimal controls and actions. Being able to visualise certain energy optimal controls and energy-saving actions increases awareness (Robic et al., 2020)	• Design

Table 4.4   Indicative simulation technologies (Khajavi et al., 2019)

Technology	Description
Various sensors	Modelica models and reduced-order models to simulate sensors
Digital twins[a]	Visualisation of the building
Virtual reality	Digital environment to interact with the building
3D scanning	Big data modelling to create a digital replica of the building
Automation Machine Learning models	Machine learning models that simulate specific building processes

[a]Chapter 6 in this book provides a detailed discussion on digital twins

Table 4.5 Commercial simulation tools

Commercial tool	Description
Vabi Apps (VABI, 2022)	Software application released by Vabi Software. It works with Autodesk Revit and is a Building Information Model (BIM) extension that incorporates building performance estimations while providing monitoring design iterations and decision support
Sefaira Architecture (Sefaira, 2022)	Software application released by Sefaira. It is a tool that is appropriate for the initial analysis of a building for efficient building performance. Sefaira implies annual simulations for the building's conceptual and schematic design
Green Building Studio (GBS, 2022)	Software application released by Autodesk. It is a feasible cloud-based application that runs building performance simulations for energy efficiency optimisation and carbon neutrality
OpenStudio (Open Studio, 2022)	OpenStudio is a software application released by the National Renewable Energy Laboratory. It is a cross-platform application that implements a collection of many tools to integrate the building's energy modelling exploiting EnergyPlus and daylight analysis
IES Virtual Environment for Architects (IESVE, 2022)	Software application released by Integrated Environmental Solutions. It is a suite of integrated tools for the design and retrofit of buildings. It supports both energy and performance simulation for the design process across the entire building life cycle
AcousticCalc—HVAC (Acousticcalc, 2022)	It is a tool for simulation of the acoustical analysis and noise prediction of the HVAC systems. ASHRAE algorithms and standards are incorporated into the tool
BuildSimHub (Buildsim, 2022)	An application for helping stakeholders to meet their design goals. It supports applications like energy model debugging and improvement and also free cloud-based energy simulations
IDA Indoor Climate and Energy (IDA, 2022)	A simulation application that models the entire building, building systems, and controllers while reducing energy consumption and providing the ideal comfort levels for the occupants
BuildingOS (BuildingOS, 2022)	A cloud application for facilities management supporting all sizes of buildings and sectors; it simulates building performance
THERM (Therm, 2022)	A software application released by Lawrence Berkeley National Laboratory (LBNL) that simulates heat transfer
EnergyPlus (EnergyPlus, 2022)	A whole building energy simulation tool. It simulates energy consumption but also supports simulations for HVAC systems and plug and process loads and water use in buildings

challenges were echoed and expanded more recently by Hong et al. (2018). The ten challenges identified by Hong et al. (2018) cover the full building life cycle and have been updated to include zero-net-energy (ZNE) and grid-responsive buildings, as well as urban-scale building energy modelling (see Table 4.6).

Table 4.6 Ten challenges of building performance simulation

Challenge	Description
Addressing the building performance gap	Previous studies have identified a significant performance gap between designed and actual energy performance of commercial and residential buildings due to mismatches in energy calculation methodologies for modelling purposes and measurements from real buildings. These gaps may be caused by miscommunication between stakeholders, inadequate quality control, and other human- and building-related factors
Modelling human-building interactions	Building occupants (and indeed owners and their agents) interact with indoor environments through their presence and adaptive actions to achieve and maintain a desirable environment. This is determined by their personal preferences and environmental attitudes, social interactions, cultural context, and indeed the physical environment and associated control options. Accurately modelling human-building interaction is an extremely difficult challenge in terms of both sourcing accurate representative data and modelling such data for different scenarios
Energy model calibration	Energy performance gaps can be addressed by calibrating building energy models with observed energy use data. While such model calibration has numerous benefits, access to such observable/ measured data is limited and the resources to collect a wider range of data is significant
Modelling operational faults in buildings	Operational faults are common; however, modelling operational faults is challenging due to the complexity (and often interdependence) and dynamic nature of faults, as well as the availability of relevant data
Modelling building operations, controls, and retrofits	As most energy is consumed during the operational phase of a building, it is unsurprising that simulation can offer significant benefits in terms of energy retrofit analysis, real-time optimisation, and control and fault detection and diagnosis. However, modelling operational faults faces distinct challenges not least most buildings lack detailed physics-based energy models or have limited current data available for generating models and training data-driven or reduced-order models. Furthermore, building owners and managers are less likely to have the IT infrastructure or BPS expertise and technical resources for supporting such processes

(continued)

Table 4.6 (continued)

Challenge	Description
Zero-net-energy (ZNE) and grid-responsive buildings	ZNE buildings typically use highly integrated design approaches and dynamic controls between all systems to optimise energy performance. Modelling passive and advanced interactive control strategies is a significant challenge. Grid-responsive buildings have similar challenges, but modelling complexity is exacerbated by differences in energy costs and the need to respond to changing conditions in near real-time which requires not only modelling the building's energy consumption but simulating renewable energy generation and utility grids, which are prone to temporal and spatial fidelity issues
Urban-scale building energy modelling	Urban-scale building energy modelling presents unique challenges in that simulations need to represent urban micro-climates that are impacted by a wide range of local factors. Furthermore, the scale and complexity of urban areas, including both buildings and the public realm, require significant high-performance computing infrastructure, resources, and tools
Evaluating the energy-saving potential of building technologies at the national or regional scale	Similar to the issues presented by urban-scale energy modelling, national- and regional-scale building modelling face discrete modelling and data capture challenges due to the large scale of the modelled phenomena over time and space
Modelling energy-efficient technology adoption	The rate of technology adoption, use, and continuance are important factors in energy simulation. Historically observed data sets play a key role in modelling, yet this is an underserved and under-research area of building performance simulation
Integrated modelling and simulation	A wide range of systems are used across a building life cycle. Greater focus on integration activities is needed across four dimensions: (1) data, (2) domain, (3) tool, and (4) workflow

Adapted from Hong et al. (2018)

While the BPS challenges Attia (2010) and Hong et al. (2018) present at a high level are not insignificant, at a more granular and operational level, the selection of a BPS tool is also not without challenges. In addition to the level of accuracy and detail, usability and information management, data exchange capacity, database support, interoperability with building modelling, and integration of building design process, Solmaz (2019) highlights the speed (time to implement), cost, and ease of use of BPS tools as significant issues. Given the centrality of BIM in the building and deep renovation[2] process, it is important to highlight specific challenges noted by many studies with respect to the integration of simulations and specifically energy simulations in BIM (Østergård et al., 2016; Hong et al., 2018; Kamel & Memari, 2019; Solmaz, 2019). Challenges include interactions between components, file-related interoperability issues at the file and syntax level, visualisation level, and semantic level, different calculation methods, attribute support, missing data, and data loss between systems (Kamel & Memari, 2019).

4.5   CONCLUSION

Building simulation can provide valuable insights across all the stages of the building and deep renovation life cycle. It may be used for many use cases as it is adaptable to various inputs and it can be deployed based on different needs. Furthermore, there is a plethora of tools and technologies that may support the simulation process. Nevertheless, simulation technology is evolving rapidly with advancements in simulation techniques, software, and hardware. Moreover, most of the simulation applications or auxiliary tools (e.g., BIM and DT) have significant data demands, leading to higher demand for storage and computational resources. As a result, new big data handling solutions must be developed to support the simulation process. In the future, building simulation will undoubtedly be a significant element in the whole cycle of the building; however, both integration and interoperability remain significant challenges that should not be underestimated.

REFERENCES

acousticcalc. (2022). http://www.acousticcalc.com/

Alor-Hernández, G., & Valencia-García, R. (Eds.). (2017). *Current trends on knowledge-based systems*. Springer International Publishing.

Attia, S. (2010). "Building performance simulation tools: selection criteria and user survey.".

Attia, S., Hamdy, M., O'Brien, W., & Carlucci, S. (2013). Assessing gaps and needs for integrating building performance optimization tools in net zero energy buildings design. *Energy and Buildings, 60*, 110–124.

Bermeo-Ayerbe, M. A., Ocampo-Martinez, C., & Diaz-Rozo, J. (2022). Data-driven energy prediction modeling for both energy efficiency and maintenance in smart manufacturing systems. *Energy, 238*, 121691.

---

[2] Chapter 1 in this book provides a detailed definition of deep renovation.

Bramstoft, R., Alonso, A. P., Karlsson, K., Kofoed-Wiuff, A., & Münster, M. (2018). STREAM—An energy scenario modelling tool. *Energy Strategy Reviews, 21*, 62–70.

buildingos. (2022). https://atrius.com/welcome-buildingos/

buildsim. (2022). https://www.buildsim.io/

Dimara, A., Anagnostopoulos, C. N., Krinidis, S., & Tzovaras, D. (2021, January). Benchmarking of regression algorithms for simulating the building's energy. *In 2021 IEEE 11th Annual Computing and Communication Workshop and Conference* (CCWC) (pp. 94–100). IEEE.

EnergyPlus. (2022). https://energyplus.net/.

Fanti, M. P., Mangini, A. M., & Roccotelli, M. (2018). A simulation and control model for building energy management. *Control Engineering Practice, 72*, 192–205.

Fernandez-Antolin, M. M., del Río, J. M., & González-Lezcano, R. A. (2022). Building performance simulations and architects against climate change and energy resource scarcity. *Earth, 3*(1), 31–44.

GBS. (2022). https://gbs.autodesk.com/GBS/

GhaffarianHoseini, A., Zhang, T., Nwadigo, O., GhaffarianHoseini, A., Naismith, N., Tookey, J., & Raahemifar, K. (2017). Application of nD BIM Integrated Knowledge-based Building Management System (BIM-IKBMS) for inspecting post-construction energy efficiency. *Renewable and Sustainable Energy Reviews, 72*, 935–949.

Hong, T., Langevin, J., & Sun, K. (2018, October). Building simulation: Ten challenges. In *Building simulation* (vol. 11, no. 5, pp. 871–898). Springer.

IDA. (2022). https://www.equa.se/en/ida-ice

IESVE. (2022). https://www.iesve.com/

Kamel, E., & Memari, A. M. (2019). Review of BIM's application in energy simulation: Tools, issues, and solutions. *Automation in Construction, 97*, 164–180.

Kampelis, N., Papayiannis, G. I., Kolokotsa, D., Galanis, G. N., Isidori, D., Cristalli, C., & Yannacopoulos, A. N. (2020). An integrated energy simulation model for buildings. *Energies, 13*(5), 1170.

Khajavi, S. H., Motlagh, N. H., Jaribion, A., Werner, L. C., & Holmström, J. (2019). Digital twin: Vision, benefits, boundaries, and creation for buildings. *IEEE Access, 7*, 147406–147419.

Laghmich, N., Romani, Z., Lapisa, R., & Draoui, A. (2022, January). Numerical analysis of horizontal temperature distribution in large buildings by thermo-aeraulic zonal approach. In *Building simulation* (vol. 15, no. 1, pp. 99–115). Tsinghua University Press.

Liu, M., Fang, S., Dong, H., & Xu, C. (2021). Review of digital twin about concepts, technologies, and industrial applications. *Journal of Manufacturing Systems, 58*, 346–361.

Loonen, R. C., Favoino, F., Hensen, J. L., & Overend, M. (2017). Review of current status, requirements and opportunities for building performance simulation of adaptive facades. *Journal of Building Performance Simulation, 10*(2), 205–223.

Mahdavi, A. (2020). In the matter of simulation and buildings: Some critical reflections. *Journal of Building Performance Simulation, 13*(1), 26–33.

Martin, Aleysha K., et al. (2022). "Co-production of a transdisciplinary assessment by researchers and healthcare professionals: a case study." Public Health Res Pract 10.

Open Studio. (2022). https://openstudio.net/

Østergård, T., Jensen, R. L., & Maagaard, S. E. (2016). Building simulations supporting decision making in early design—A review. *Renewable and Sustainable Energy Reviews, 61*, 187–201.

Oucquier, A., Robert, S., Suard, F., Stephan, L., & Jay, A. (2013). State of the art in building modeling and energy performances prediction: A review. *Renewable and Sustainable Energy Reviews, 23*, 272–288.

Puttur, U., Ahmadi, M., Ahmadi, B., & Bigham, S. (2022). A novel lung-inspired 3D-printed desiccant-coated heat exchanger for high-performance humidity management in buildings. *Energy Conversion and Management, 252*, 115074.

Robic, F., et al. (2020). Implementation and fine-tuning of the Big Bang-Big Crunch optimisation method for use in passive building design. *Building and Environment, 173*, 106731.

Sefaira. (2022). http://www.tenlinks.com/news/vabi-revit-apps-available-in-autodesk-exchange-apps-store/

Shen, R., Jiao, Z., Parker, T., Sun, Y., & Wang, Q. (2020). Recent application of Computational Fluid Dynamics (CFD) in process safety and loss prevention: A review. *Journal of Loss Prevention in the Process Industries, 67*, 104252.

Solmaz, A. S. (2019). A critical review on building performance simulation tools. *Alam Cipta, 12*(2), 7–21.

Tan, Y., Peng, J., Luo, Y., Gao, J., Luo, Z., Wang, M., & Curcija, D. C. (2022). Parametric study of venetian blinds for energy performance evaluation and classification in residential buildings. *Energy, 239*, 122266.

therm. (2022). https://windows.lbl.gov/tools/therm/software-download

Tian, Z. (2018b). Towards adoption of building energy simulation and optimization for passive building design: A survey and a review. *Energy and Buildings, 158*, 1306–1316.

VABI. (2022). http://www.tenlinks.com/news/vabi-revit-apps-available-in-autodesk-exchange-apps-store/

Vera, S. (2018). A critical review of heat and mass transfer in vegetative roof models used in building energy and urban environment simulation tools. *Applied Energy, 232*, 752–764.

Yu, Y. A. (2019). A review of the development of airflow models used in building load calculation and energy simulation. *Building Simulation, 12*(3).

CHAPTER 5

Big Data and Analytics in the Deep Renovation Life Cycle

Paraskevas Koukaras, Stelios Krinidis,
Dimosthenis Ioannidis, Christos Tjortjis,
and Dimitrios Tzovaras

Abstract The rising volume of heterogeneous data accessible at various phases of the construction process has had a significant impact on the construction industry. The availability of data is especially advantageous in the context of deep renovation, where it may significantly accelerate the decision-making process for building stock retrofit. This chapter covers Big Data and analytics in the context of deep renovation and shows how Machine Learning and Artificial Intelligence have affected the various phases of the deep renovation life cycle. It presents a review of the literature on Big Data and deep renovation and discusses a series of use cases,

P. Koukaras (✉) • C. Tjortjis
Information Technologies Institute, Centre for Research & Technology Hellas
(CERTH), Thessaloniki, Greece

School of Science and Technology, International Hellenic University,
Thessaloniki, Greece
e-mail: p.koukaras@iti.gr; c.tjortjis@iti.gr

© The Author(s) 2023   69
T. Lynn et al. (eds.), *Disrupting Buildings*, Palgrave Studies in
Digital Business & Enabling Technologies,
https://doi.org/10.1007/978-3-031-32309-6_5

applications, advantages, and benefits as well as challenges and barriers. Finally, Big Data and deep renovation prospects are discussed, including future potential developments and guidelines.

Keywords Big Data • Deep Renovation • Artificial Intelligence • Machine Learning • Energy

5.1 INTRODUCTION

As in many industries, the construction sector has been impacted and somewhat changed by the growing volume of heterogeneous data available at different stages of the construction process. This trend is expected to continue as technologies such as sensors and the Internet of Things (IoT) become more and more accessible and commoditised. The availability of data is particularly useful in the context of Deep Renovation (DR)[1] where it can dramatically accelerate the decision-making for building stock retrofit. This chapter defines Big Data (BD) and analytics in the context of DR and describes how the use of BD and advanced analytics such as Machine Learning (ML) and Artificial Intelligence (AI) may impact different stages of the DR life cycle.

The remainder of this chapter is organised as follows. Section 5.2 provides an overview of BD and different types of analytics. Section 5.3 presents a series of use cases and applications of BD in construction. Section 5.4 describes how BD can be used in the DR space. Section 5.5

---

[1] Chapter 1 in this book provides a detailed definition of deep renovation.

S. Krinidis
Information Technologies Institute, Centre for Research & Technology Hellas (CERTH), Thessaloniki, Greece

Department of Management Science and Technology, International Hellenic University, Kavala, Greece
e-mail: krinidis@mst.ihu.gr

D. Ioannidis • D. Tzovaras
Information Technologies Institute, Centre for Research & Technology Hellas (CERTH), Thessaloniki, Greece
e-mail: djoannid@iti.gr; Dimitrios.Tzovaras@iti.gr

discusses the advantages and benefits of BD in the context of DR. Section 5.6 outlines challenges and barriers related to the adoption and use of BD in construction generally and in DR more specifically. Section 5.7 presents potential future developments and, finally, Sect. 5.8 contains some concluding remarks.

5.2   BIG DATA ANALYTICS

BD analytics deals with large, heterogeneous data sets from various sources. Data-driven decision-making entails finding trends, patterns, and correlations in data. In order to do that, different types of BD analytics can be implemented which can be classified under four main categories— that is, descriptive, diagnostic, predictive, and prescriptive analytics. Data mining, cleansing, integration, and visualisation enable data analytics in various domains and change and/or improve DR processes, delivering commercial and societal benefits (Rajaraman, 2016; Koukaras & Tjortjis, 2019; Kousis & Tjortjis, 2021).

Descriptive analytics is a popular way for organisations to analyse current and past trends and operational performance. It is the initial stage in interpreting raw data by applying relatively basic statistics and creating sample and measurement statements.

Diagnostic analytics is a type of BD analytics used to evaluate data and content. This form of analytics typically answers questions like 'why did something happen?' and therefore aims to explain the causes behind particular results.

Predictive analytics involves estimating outcomes using data insights. It typically employs ML and statistical modelling to predict the most likely outcomes.

Prescriptive analytics is built on the insights gained from descriptive, diagnostic, and predictive analytics to optimise operational processes using simulations and related tools (see Chap. 4 for more details). It uses statistics and data modelling to assist organisations understand and predict the market or environment. It helps individuals define priorities and recognise what might lead to financial or other types of rewards.

5.3   USE CASES AND APPLICATIONS OF BIG DATA IN CONSTRUCTION

BD analytics is backed by BD engineering, which has significant construction applications. BD engineering involves Building Information Modelling (BIM)[2] to enhance project management (Huang, 2021), building design and monitoring performance (Loyola, 2018), safety, energy management, decision-making design frameworks, resource management (Ismail et al., 2018), quality management, waste management (Wang et al., 2018), and more (see Chap. 3 for a more detailed discussion). Moreover, BD platforms that perform BD analytics in construction are essential to BD engineering and may be classified as Horizontally Scaling Platforms (HSPs) and Vertically Scaling Platforms (VSPs). HSPs use several servers by spreading processes and adding additional devices, while VSPs scale by updating the server's hardware. Waste management (Bilal et al., 2016b), profitability performance measurement (Bilal et al., 2019), smart road construction, and others (Sharif et al., 2017) typically employ HSPs, while VSPs have been mostly used in construction (Curtis, 2020) and transportation (Shtern et al., 2014).

Furthermore, deep learning–based flood detection and damage assessment (Munawar et al., 2021), project delay risk prediction (Gondia et al., 2020), construction site safety (Tixier et al., 2016), construction site monitoring (Rahimian et al., 2020), and neural network models to predict concrete qualities (Maqsoom et al., 2021) are a few instances of AI and ML in construction.

5.4   BIG DATA AND DEEP RENOVATION

The fundamental components of BD engineering include both distributed and parallel processing. BD analytics has been used in the construction industry for a variety of purposes, including waste management (Lu et al., 2016), management of prefabricated building projects (Han & Wang, 2017), profitability studies, and other construction management applications (Bilal et al., 2019).

BD in construction uses AI and ML for revitalising sustainable architecture, energy-efficient building design, and minimising environmental and climatic consequences. Recent advancements in internet speed,

---

[2] Chapter 3 in this book provides a detailed discussion on BIM.

accessibility, processing cost, and data storage cost make BD a vital AI supplement (Mehmood et al., 2019).

In recent years, AI has contributed significantly to improving learning-based decision-making. Its use in building design and engineering along with BIM is offering new options for DR utilising BD since very large volumes of construction-related data are available. DR is one of the main drivers for Greenhouse Gas (GHG) emission reduction in cities (Avramidou & Tjortjis, 2021) and along with ML and AI introduces new design potentials, constraints, and solutions. Overall, BIM and Industry Foundation Classes (IFC) improve DR's decision-making and the energy efficiency of retrofitted buildings (Mulero-Palencia et al., 2021).

Nowadays, DR should aim to harness the maximum economic energy efficiency potential of construction activities at a large scale, thus utilising BD for construction purposes. It should also concentrate on improvements of the building shell of existing structures, leading to extremely high-energy performance. Nonetheless, residential efficiency improvements or criteria (e.g. shell upgrades or Heating, Ventilation and Air Conditioning (HVAC) and hot water system upgrades) vary by climate (Cluett & Amann, 2014).

Despite the European Union's (EU) energy efficiency targets and renovation actions such as aesthetic improvement of the building outer façade, increased thermal comfort and energy efficiency, and CO2 emission minimisation, the construction industry has not yet adopted large-scale standardised retrofitting techniques that would involve BD analytics in construction (Glumac et al., 2013). Most renovation options include external/internal insulation, air tightening the transparent and opaque building envelope, roof conversion, solar panels, heat recovery, and more efficient HVAC systems. Conventional energy retrofits focus on single system upgrades, such as façade, lighting, and HVAC equipment, without considering integrated renovation options.

5.5 ADVANTAGES AND BENEFITS

Literature suggests a number of opportunities for BD adaptation in the DR context (Bilal et al., 2016a):

1. *Generative design.* The idea is to automate the development of several design models based on specific objectives such as functional requirements, material type, manufacturing process, performance standards, and cost limitation. Such tools use advanced algorithms to develop design solutions that fulfil design criteria. Designers eval-

uate the designs' performance and are able to change design objectives and restrictions until they are satisfied.

2. *Clash detection and resolution.* BIM models should identify design incompatibilities. For effective project management, this step should come before construction. Traditional paper-based procedures, which are less efficient and accurate in identifying design issues, are being replaced by BIM-enabled automated techniques. Design conflict detection involves time-consuming non-trivial design exploration strategies. BD technology may improve knowledge representation and processing via distributed and parallel computing.

3. *Performance prediction models.* These models employ a vast number of variables and their combinations, which affect each other and overall model performance. They are implemented utilising basic statistical approaches or more complex computational methods such as artificial neural networks. Therefore, these systems involve a large number of variables, something that is computationally intensive, time-consuming, and difficult for existing technologies to perform in real-time. BD technologies may improve real-time processing, model creation, and visualisation.

4. *Visual analytics.* Analytical issues may be categorised as the ones that require logical solutions and the ones that require heuristic solutions. The first can be automated, whereas the latter are tackled by proper visualisation. Effective visualisation requires human expertise, imagination, and intuition; thus, human knowledge works well with smaller data sets but not with high-dimensional data sets. Visual analytics combines automated reasoning and graphical representations to address complicated analytical issues and require BD to visualise data for enabling personal viewpoints and interactive data exploration.

5. *Social networking.* The majority of construction sector issues are communication-related. Social media can pose as a fascinating development that might help businesses promote good communication among project teams. Communication via social media is a tendency that has been steadily infiltrating the business sector. The next application areas might include social networking platforms for sharing updated project information, as well as other initiatives for conveying the best sustainability strategies/practices. Yet, strong frameworks need to be conceived to capture all valuable social interactions in BIM forms, from initial design to the final model. Since social media data are likely to be varied, rapid, and massive, BD may be used to construct novel domain applications to promote stakeholder productivity.

6. *Personalised services.* Such services emphasise on adapting facilities to user preferences. Users control how services are utilised and these systems adapt to user behaviour. They consider both human and automated input. Therefore, personalisation solutions monitor the surroundings for occurrences of interest, creating vast amounts of data. BD technologies can analyse these data streams in real-time to create actionable insights for nearly instant adaption. To do so, contemporary buildings need BD-enabled platforms with a uniform interface to facilitate such personalisation services.

Other advantages/benefits of BD and analytics in construction include:

1. *Improvements in construction efficiency.* By delivering clear, comprehensible data and detecting possible structural flaws before they occur, data analytics technology aims to cut construction time and material costs. This enables project managers to make faster and more informed choices, thus reducing human errors (Lynn et al., 2021).
2. *Environmental impact reduction.* Integrating BD that is actually historical project construction data can be blended into BIM technology to precisely estimate the materials and energy required for upcoming projects. This cuts down on unnecessary building waste and enables planners to explore more options for energy-efficient solutions when feasible (Androutsopoulos et al., 2020).

5.6 CHALLENGES AND BARRIERS

In recent years, energy efficiency has become one of the EU's top priorities (Koukaras et al., 2021a). Some generic barriers of BD in DR are (Lynn et al., 2021) as follows:

1. *Human.* Several variables may impede the approval, support, and adoption of energy-efficient behaviours, technologies, and initiatives. Social norms, behavioural patterns, inability to use new technologies, lack of information on energy consumption and energy-saving opportunities, and more are all barriers. Moreover, education, age, and family composition affect the adoption of energy-efficient equipment. All these underscore the necessity of adjusting communication to various groups and educating construction experts for adopting and using BD analytics for analysing these data to elevate DR.

2. *Technological integration.* DR when supported by BD involves multiple domains, stakeholders, and technologies. Interoperability improves communication, coordination, cooperation, collaboration, and distribution in DR projects. This causes interoperability concerns that hinder data flows and value creation when BD is involved. Linking data throughout a restoration project's lifespan offers a variety of obstacles such as detecting and reconciling disparate schemas and object representations, incompatibilities across data sources, incompatible levels of abstraction, and data quality concerns.

3. *Organisational.* DR demands senior management commitment with interdisciplinary skills, time, expenditure, skilled personnel, and appropriate technical infrastructure. The absence of appropriately qualified energy efficiency specialists, data scientists, and construction workers in the right selection and installation for integrating new constructional technologies is a major obstacle for DR especially when supported by BD analytics. Project delays and interruptions, sub-optimal energy efficiency, and failure to achieve expected cost reductions as well as high initial investment costs, finance availability, and payback time are also obstacles.

More barriers related specifically to data, that is, BD, applicable in the context of DR are (Bilal et al., 2016a): (a) data security, privacy, and protection, (b) data quality of construction industry data sets, and (c) fast and reliable internet connectivity for BD applications.

In addition, other challenges related to BD handling in construction are (Yousif et al., 2021): (a) inefficient BD experts/data collectors, analysts, and presenters along with the dynamic nature (e.g. online data streams) of BD databases; (b) high expenditure in BD infrastructure/experts, which will prevent enterprises from adopting BD technologies; (c) governments and corporations, which avoid sharing important data with the world, thus forcing data protection policies to be established.

Furthermore, another study specifically looks at BD for energy efficiency in building and notes data access challenges (Marinakis, 2020).

Finally, possible barriers are also related to social and environmental aspects in the context of BD. Aside from reducing energy consumption (Koukaras et al., 2021b), building renovations are typically motivated by issues such as structural repairs (D'Agostino et al., 2017). Buildings utilise 38% of EU energy and produce 36% of CO2. For example, the Dutch non-profit building stock's DR ratings for 2010–2014 were based on the energy performance of 850,000 homes. The data were obtained from a system that monitored 60% of the sector's buildings. Despite renovations, the dwellings' energy efficiency did not alter much (Filippidou et al., 2017).

5.7 FUTURE DEVELOPMENTS

BD integration potentially benefits construction companies and all the other stakeholders involved at different stages of the DR life cycle. Using BD for business and environmental sustainability offers construction companies major prospects. BD may help the building sector overcome present hurdles. Using historical and current project data may assist in fostering long-term infrastructure. BD in construction helps prevent errors and yield better construction outcomes.

Future studies could investigate the integrated data that will be utilised for worldwide commercialisation of BD analytics for DR. This involves developing web/mobile applications that can be linked to BD integration systems to show real-time data analytics at a low cost, as well as work on the data-gathering process in the construction fields (Yousif et al., 2021).

Furthermore, future work foresees aspects related to (a) construction waste simulation tools, (b) BD analytics that enables linked building data platforms, (c) BD-driven BIM systems for construction progress monitoring, and (d) BD for design with data (Bilal et al., 2016a).

Future construction research will depend largely on BD since it can help develop better infrastructure and building designs. Construction must automate and integrate technologies to make BD utilisation simple and easy. BD technologies, BIM, and Computer-Aided Design (CAD) cannot be used without proper support and integration. The building industry's future rests on steadily improving the current conditions (Gbadamosi et al., 2020).

Finally, BD is essential to future building DR projects and data are essential for establishing training models and facilitating construction in general. Future improvements in this area will involve additional algorithms and models that depend on BD for reliable training.

5.8 CONCLUSION

The objective of this chapter was to expose the reader to the concepts of BD and DR. Technologies, such as prefabricated exteriors, ICT-support for Building Management Systems (BMS), incorporation of Renewable

Energy Systems (RES), BIM and building performance simulation models, and high-tech HVAC systems, are just not enough for reaching EU climate change policy goals by 2050. There are still open issues for future innovation in order to conceive effective policies and suggestions for DR implementations (D'Oca et al., 2018). Thus, BD analytics can be employed for the construction sector and more specifically in the context of DR passing on the discussed advantages and benefits.

Nonetheless, the building sector has not yet fully embraced BD. The fast rise of this technology over the past two decades has increased the number of models and platforms for digitising diverse areas. The literature reveals several resources and platforms that may be used for construction management. Yet, currently there is poor adoption in DR and the building business must use and commercialise BD.

Future developments will benefit from internet tools and technologies that allow infrastructure modelling and CAD. These relate to the implementation of efficient energy measures, any prospects for climate change mitigation, and better management for thermal comfort in the context of BD and DR. Simple and inexpensive renovations frequently miss the opportunity to save more energy at a reduced cost. Any DR initiative should include many locations with different building, regulatory, market, and climatic conditions, thus involving BD.

The importance of BD and analytics for enhancing DR was highlighted using representative paradigms from the recent state-of-the-art. Social, economic, and environmental perspectives were also taken into account. In order to make the most out of the large amount of information accessible in the current BD environment, new analytical skills for DR must be developed.

References

Androutsopoulos, A., Geissler, S., Charalambides, A. G., Escudero, C. J., Kyriacou, O., & Petran, H. (2020). Mapping the deep renovation possibilities of European buildings. *IOP Conference Series: Earth and Environmental Science,* *410*(1), 12056. https://doi.org/10.1088/1755-1315/410/1/012056

Avramidou, A., & Tjortjis, C. (2021). In I. Maglogiannis, J. Macintyre, & L. Iliadis (Eds.), *Predicting CO$_2$ emissions for buildings using regression and classification BT—Artificial Intelligence applications and innovations* (pp. 543–554). Springer International Publishing.

Bilal, M., Oyedele, L. O., Akinade, O. O., Ajayi, S. O., Alaka, H. A., Owolabi, H. A., Qadir, J., Pasha, M., & Bello, S. A. (2016b). Big Data architecture for

Construction Waste Analytics (CWA): A conceptual framework. *Journal of Building Engineering, 6*, 144–156.

Bilal, M., Oyedele, L. O., Kusimo, H. O., Owolabi, H. A., Akanbi, L. A., Ajayi, A. O., Akinade, O. O., & Delgado, J. M. D. (2019). Investigating profitability performance of construction projects using Big Data: A project analytics approach. *Journal of Building Engineering, 26*, 100850.

Bilal, M., Oyedele, L. O., Qadir, J., Munir, K., Ajayi, S. O., Akinade, O. O., Owolabi, H. A., Alaka, H. A., & Pasha, M. (2016a). Big Data in the construction industry: A review of present status, opportunities, and future trends. *Advanced Engineering Informatics, 30*(3), 500–521. https://doi.org/10.1016/j.aei.2016.07.001

Cluett, R., & Amann, J. (2014). *Residential deep energy retrofits*. March, 64. www.aceee.org

Curtis, C. (2020). Architecture at scale: Reimagining one-off projects as building platforms. *Architectural Design, 90*(2), 96–103.

D'Agostino, D., Zangheri, P., & Castellazzi, L. (2017). Towards nearly zero energy buildings in Europe: A focus on retrofit in non-residential buildings. *Energies, 10*(1), 117. https://doi.org/10.3390/en10010117

D'Oca, S., Ferrante, A., Ferrer, C., Pernetti, R., Gralka, A., Sebastian, R., & Op 't Veld, P. (2018). Technical, financial, and social barriers and challenges in deep building renovation: Integration of lessons learned from the H2020 cluster projects. *Buildings, 8*(12). https://doi.org/10.3390/buildings8120174

Filippidou, F., Nieboer, N., & Visscher, H. (2017). Are we moving fast enough? The energy renovation rate of the Dutch non-profit housing using the national energy labelling database. *Energy Policy, 109*, 488–498. https://doi.org/10.1016/j.enpol.2017.07.025

Gbadamosi, A.-Q., Oyedele, L., Mahamadu, A.-M., Kusimo, H., Bilal, M., Delgado, J. M. D., & Muhammed-Yakubu, N. (2020). Big Data for Design Options Repository: Towards a DFMA approach for offsite construction. *Automation in Construction, 120*, 103388.

Glumac, B., Reuvekamp, S., Han, Q., & Schaefer, W. (2013). Tenant participation in sustainable renovation projects: Using AHP and case studies. *Journal of Energy Technologies and Policy—Special Issue for International Conference on Energy, Environment and Sustainable Economy (EESE 2013), 3*(11), 16–26.

Gondia, A., Siam, A., El-Dakhakhni, W., & Nassar, A. H. (2020). Machine learning algorithms for construction projects delay risk prediction. *Journal of Construction Engineering and Management, 146*(1), 4019085.

Han, Z., & Wang, Y. (2017). The applied exploration of Big Data technology in prefabricated construction project management. In *ICCREM 2017* (pp. 71–78). ASCE Library.

Huang, X. (2021). Application of BIM Big Data in construction engineering cost. *Journal of Physics: Conference Series, 1865*(3), 32016.

Ismail, S. A., Bandi, S., & Maaz, Z. N. (2018). An appraisal into the potential application of Big Data in the construction industry. *International Journal of Built Environment and Sustainability, 5*(2).

Koukaras, P., Bezas, N., Gkaidatzis, P., Ioannidis, D., Tzovaras, D., & Tjortjis, C. (2021a). Introducing a novel approach in one-step ahead energy load forecasting. *Sustainable Computing: Informatics and Systems, 32*, 100616. https://doi.org/10.1016/j.suscom.2021.100616

Koukaras, P., Gkaidatzis, P., Bezas, N., Bragatto, T., Carere, F., Santori, F., Antal, M., Tjortjis, C., & Tzovaras, D. (2021b). A tri-layer optimization framework for day-ahead energy scheduling based on cost and discomfort minimization. *Energies, 14*(12), 3599. https://doi.org/10.3390/en14123599

Koukaras, P., & Tjortjis, C. (2019). Social media analytics, types and methodology. In *Machine learning paradigms* (pp. 401–427). Springer. https://doi.org/10.1007/978-3-030-15628-2_12

Kousis, A., & Tjortjis, C. (2021). Data mining algorithms for smart cities: A bibliometric analysis. *Algorithms, 14*(8), 242. https://doi.org/10.3390/a14080242

Loyola, M. (2018). Big Data in building design: A review. *Journal of Information Technology in Construction, 23*, 259–284.

Lu, W., Chen, X., Ho, D. C. W., & Wang, H. (2016). Analysis of the construction waste management performance in Hong Kong: The public and private sectors compared using Big Data. *Journal of Cleaner Production, 112*, 521–531.

Lynn, T., Rosati, P., Egli, A., Krinidis, S., Angelakoglou, K., Sougkakis, V., Tzovaras, D., Kassem, M., Greenwood, D., & Doukari, O. (2021). Rinno: Towards an open renovation platform for integrated design and delivery of deep renovation projects. *Sustainability (Switzerland), 13*(11). https://doi.org/10.3390/su13116018

Maqsoom, A., Aslam, B., Gul, M. E., Ullah, F., Kouzani, A. Z., Mahmud, M. A. P., & Nawaz, A. (2021). Using multivariate regression and ANN models to predict properties of concrete cured under hot weather. *Sustainability, 13*(18), 10164.

Marinakis, V. (2020). Big Data for energy management and energy-efficient buildings. *Energies, 13*(7). https://doi.org/10.3390/en13071555

Mehmood, M. U., Chun, D., Zeeshan, H., Jeon, G., & Chen, K. (2019). A review of the applications of artificial intelligence and Big Data to buildings for energy-efficiency and a comfortable indoor living environment. *Energy and Buildings, 202*, 109383. https://doi.org/10.1016/j.enbuild.2019.109383

Mulero-Palencia, S., Álvarez-Díaz, S., & Andrés-Chicote, M. (2021). Machine learning for the improvement of deep renovation building projects using as-built bim models. *Sustainability (Switzerland), 13*(12). https://doi.org/10.3390/su13126576

Munawar, H. S., Ullah, F., Qayyum, S., & Heravi, A. (2021). Application of deep learning on UAV-based aerial images for flood detection. *Smart Cities, 4*(3), 1220–1242.

Rahimian, F. P., Seyedzadeh, S., Oliver, S., Rodriguez, S., & Dawood, N. (2020). On-demand monitoring of construction projects through a game-like hybrid application of BIM and machine learning. *Automation in Construction, 110*, 103012.

Rajaraman, V. (2016). Big DATA analytics. *Resonance, 21*(8), 695–716. https:// doi.org/10.1007/s12045-016-0376-7

Sharif, M., Mercelis, S., Van Den Bergh, W., & Hellinckx, P. (2017). Towards real-time smart road construction: Efficient process management through the implementation of internet of things. *Proceedings of the International Conference on Big Data and Internet of Thing, 2017*, 174–180.

Shtern, M., Mian, R., Litoiu, M., Zareian, S., Abdelgawad, H., & Tizghadam, A. (2014). Towards a multi-cluster analytical engine for transportation data. *2014 International Conference on Cloud and Autonomic Computing, 2014*, 249–257.

Tixier, A. J.-P., Hallowell, M. R., Rajagopalan, B., & Bowman, D. (2016). Application of machine learning to construction injury prediction. *Automation in Construction, 69*, 102–114.

Wang, D., Fan, J., Fu, H., & Zhang, B. (2018). Research on optimization of Big Data construction engineering quality management based on RNN-LSTM. *Complexity, 2018*.

Yousif, O. S., Zakaria, R. B., Aminudin, E., Yahya, K., Mohd Sam, A. R., Singaram, L., Munikanan, V., Yahya, M. A., Wahi, N., & Shamsuddin, S. M. (2021). Review of Big Data integration in construction industry digitalization. *Frontiers in Built Environment, 7*(November), 1–13. https://doi.org/10.3389/fbuil. 2021.770496

CHAPTER 6

Digital Twins and Their Roles in Building Deep Renovation Life Cycle

Yuandong Pan, Zhiqi Hu, and Ioannis Brilakis

Abstract Digital twins have started to diffuse within architecture, engineering, construction, and operations (AECO), based on their emerging and anticipated benefits to the various stakeholders involved in the building life cycle. However, their applications are still at an early stage, and much effort is still needed to exploit their full potential. This chapter explains some key notions to help understand digital twins in AECO. It exposes the various definitions of digital twins and illustrates the basic steps and relevant methods for creating a digital twin. The chapter also provides an overview of the state-of-the-art deep learning methods for digital twins and discusses some real-life use cases. Finally, the chapter discusses the benefits and challenges associated with the adoption of digital twins.

Y. Pan
Technical University of Munich, Munich, Germany

Z. Hu (✉) • I. Brilakis
Construction Information Technology Laboratory (CIT Lab), University of Cambridge, Cambridge, UK
e-mail: zh334@cam.ac.uk

© The Author(s) 2023   83
T. Lynn et al. (eds.), *Disrupting Buildings*, Palgrave Studies in Digital Business & Enabling Technologies,
https://doi.org/10.1007/978-3-031-32309-6_6

Keywords Scanning • Digital twinning • Geometry • Deep learning

6.1   INTRODUCTION

The construction sector remains one of the least digitised sectors. Digitalisation and automation can prove particularly valuable in overcoming a number of traditional challenges in architecture, engineering, construction, and operations (AECO). First, over half of the labour time is spent waiting for materials, equipment, and instructions on how to conduct the work during the construction stage, resulting in low productivity and shrinking profit margins. Second, many construction companies have suffered from underperforming projects, which leads to cost and schedule overruns and asset's quality issues. Third, many assets are designed for functional activities. Less consideration is given to their environmental impact leading to high carbon emissions and resource wastage. Fourth, due to skill shortage, it is difficult to recruit enough construction professionals, such as supervisors, estimators, and engineers, which exacerbates the issue related to delays, asset qualities, and safety.

Digital twin (DT) is an emerging technological paradigm for achieving smart buildings, infrastructure, and cities (Tao et al., 2019). DT applications can facilitate project management in the AECO sector by increasing productivity and efficiency. From manual drawings to computer-aided design, object-oriented design, and computational design, computer power is shaping the process of assets' construction and maintenance by encoding decision-makings through machine learning and other advanced technologies. This chapter aims to provide an overview of digital twins and their applications in the context of building renovation and discuss their main advantages, benefits, challenges, and barriers to adoption. The next section presents the definition of digital twins. The following section presents the main steps for creating a digital twin. This is followed by the presentation of a series of use cases and some concluding remarks on potential future developments.

6.2   WHAT IS A DIGITAL TWIN

According to Tao et al. (2019), a DT consists of three main elements: a physical product, a virtual representation of the physical product, and the connection that links these two parts together and enables data exchange and information sharing. The physical product refers to the actual asset built in the real (physical) world, which can also be defined as physical twin (PT). It can be a residential or a commercial building, a hospital, a school, a bridge, and so on. The virtual representation refers to the digital replica of the physical asset, which can exist throughout its life cycle. This data can be accumulated over time and updated at different stages of a physical asset's lifetime. The connection that links these two parts can be considered as an information exchanger to store, link, and update all product and process information over time. A DT can serve as an information repository for storing and sharing an asset's properties throughout its life cycle (El Saddik, 2018).

According to Sacks et al. (2020), a DT is dynamic and thus can be enriched through different stages of an asset's life cycle. Figure 6.1 depicts a typical life cycle of an asset PT and its DT from the design stage, through the construction stage, to the operation stage.

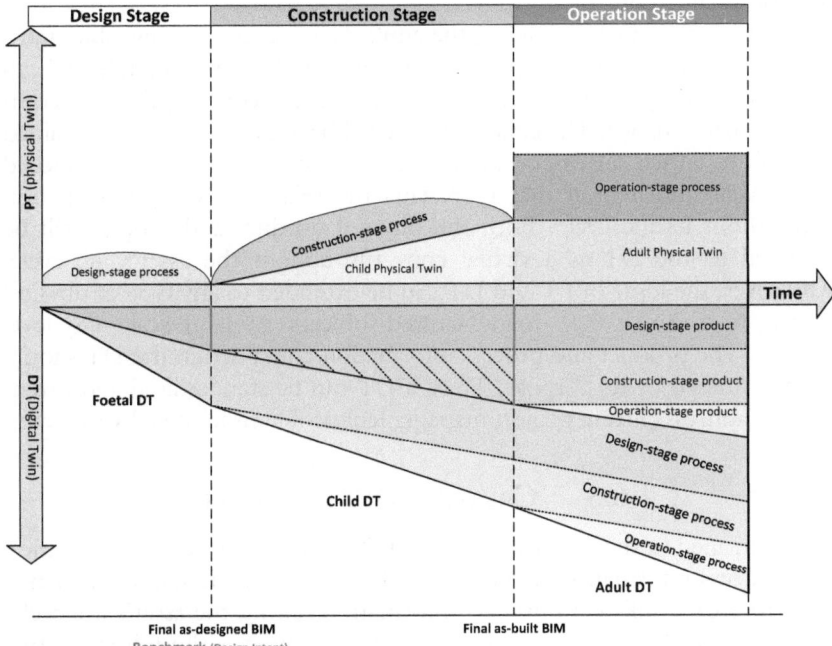

Fig. 6.1 A typical life cycle of an asset PT and its DT from the design, construction, to the operation stage

At the design stage, the asset's designers start working on the conceptual plan. The asset's foetal DT contains both product and process information, where the former refers to different as-designed building information models (BIMs).[1] Many of these models can be proposed at the beginning, but only the final client-approved design file at the end of the design stage can be marked as "Design Intent", which means it will serve as a benchmark for evaluating the construction outcomes and can be considered as a guidance for the purpose of maintenance.

At the construction stage, the child PT contains off-site prefabricated assemblies and on-site constructed components. Therefore, the child DT consists of as-built product information and as-performed process information to mirror the asset's physical status at different steps during the construction stage. It should be noted that the product information and the process information accumulate over time into the child DT until the completion of construction. Each change will be updated in the asset's child DT to reflect the as-is status and thus can facilitate progress monitoring and quality control.

Lastly, at the operation stage, the adult PT remains unchangeable status because of the completion of the construction. The asset's adult DT can support the analyses of performance, such as energy consumption and component maintenance. The collected data will be added to the as-maintained product to enrich the asset's adult DT. To conclude, an asset's DT should contain all information that represents the related physical information throughout its life cycle. Both the physical product and process will be assigned to the DT as a virtual copy throughout the asset's life cycle. Moreover, the logic of PT and DT can be extended to any type of physical entity, from small-scale manufactured objects to large-scale city-level objects. The product and process information contained in the DT should be determined by its purpose. Thus, a DT can be standardised and extensible to address current project management problems in the AECO sector.

6.3   Creating Digital Twins

As mentioned in the previous section, a DT contains product information and process information. A geometric DT (GDT) is fundamental as it is used to create links with process information during the asset's life cycle. Creating a GDT of an existing asset typically involves the following two

---

[1] Chapter 3 in this book provides a more detailed discussion on BIM.

steps: (1) capturing raw visual and spatial data in the form of RGB images and laser-scanned point clouds and (2) detecting geometric objects and relationships between objects. Step 1 of this process is significantly more automated than step 2, as shown by Agapaki and Brilakis (2021). Unfortunately, the effort and corresponding cost required to complete step 2 for most assets still represent a barrier to adoption as it may completely offset the value created by the geometric DT.

For data capturing (step 1), two major technologies are currently used to capture the geometry of an asset: laser scanning (terrestrial and mobile) and photogrammetry. The data generated should reflect the physical surfaces of objects in the real world. Due to the discrete nature of the capturing techniques, the data provided by scanners is also discrete. Laser scanners generate point clouds that are sets of points in a 3D space. Each point is defined by three coordinates and additional information depending on the device used, which could be intensity, normality, and colour information, among others.

As for step 2, detecting geometric objects and their geometric relationships is still a time-consuming manual task. Lu et al. (2019), for example, scanned ten different road bridges and estimated that approximately 28 hours of work are required, on average, for the as-is modelling in contrast to 2.82 hours for data capturing. A number of leading 3D CAD companies (Autodesk, Bentley, ClearEdge3D, etc.) have developed software products that provide a variety of 3D modelling features which enable modelling from point cloud data. Agapaki et al. (2018) suggest that 64% of man-hour savings can be achieved by using state-of-the-art software supporting a semi-automated modelling process. However, 2382 man-hours are still needed to model, for example, a small petrochemical plant with 240,687 objects and 53,834 pipes.

In order to reduce the human effort in creating a GDT, researchers have proposed a number of alternative approaches mostly focused on structural elements. Sanchez and Zakhor (2012) proposed a method that applies principal component analysis (PCA) and random sample consensus (RANSAC) to find relatively large-scale architectural structures, such as ceilings and floors. Monszpart et al. (2015) extracted planar structures in a point cloud that follows regularity constraints. They applied this approach in different scenarios, such as urban scenes, as well as the exterior and interior of buildings. Oesau et al. (2014) used horizontal slicing and then volumetric-cell labelling method. The volumetric cells are formulated as energy minimisation and solved by the graph-cut method. Xiao and

Furukawa (2014) proposed a method called "inverse constructive solid geometry (CSG)" which detects planar surfaces and subsequently fits the cuboid primitives to the point cloud. Ochmann et al. (2016) proposed a method that explicitly represents buildings as interconnected volumetric wall elements. They determined the optimal room and wall layout by graph-cut-based multi-label energy minimisation. A method named void-growing method by Pan et al. (2021) aims to extract void room spaces in the point cloud firstly and subsequently extract 3D models of different objects.

Other approaches leverage prior knowledge to reconstruct walls and rooms. Stambler and Huber (2015), for example, proposed the concept of enclosure reasoning that defines rooms as cycles of walls enclosing free interior space. Region growing is then applied to segment the point clouds, and simulated annealing is used to optimise rooms and walls. Tran et al. (2019) proposed a method called shape grammar to model indoor environments. They created 3D parametric models by placing cuboids into point clouds and classifying them into elements and spaces. The wall candidates are obtained from pairs of adjacent peaks in the histogram of point coordinates. Hu et al. (2022) provide a more in-depth review of this literature.

Deep learning (DL) is also widely applied to extract semantic information from spatial and visual data. VoxNet is proposed by Maturana and Scherer (2015) to detect classes of objects from point cloud data. It aims to predict a class label for the input. Volumetric grids representing the spatial occupancy are calculated first and then applied to 3D CNNs. Qi et al. (2017a) instead proposed the first neural network architecture, PointNet, designed for 3D deep learning in the point cloud. PointNet takes the point cloud as input and predicts labels for the entire input (point cloud classification) or labels for each point (point cloud segmentation). An improved version of the PointNet architecture called PointNet++ has then been presented by Qi et al. (2017b) and claims to provide better performance by considering spatial information of points in the point sets. These DL methods have been adopted in the AECO sector to facilitate GDT construction (Agapaki & Brilakis, 2020; Perez-Perez et al., 2021).

In summary, current approaches are still not fully automated, which means they still require human effort in the process of reconstruction. Their performance, especially when applied to a point cloud with high occlusions, would decrease due to the geometric occlusion of furniture. On the other hand, DL is an efficient and powerful tool that can be used to extract semantic information from the point cloud, but the lack of labelled data sets in the AECO domain causes difficulties with regard to training which in turn affects models' performance. In addition to this, the overall prediction performance differs significantly across categories, which makes it really hard to create a detailed GDT representing the current state of an asset when only considering the output of the DL methods.

6.4 DIGITAL TWIN USE CASES

There are several use cases of DT in the construction sector, including construction progress monitoring, facilities management and operation, asset condition monitoring, sustainable development, and more. DT can provide reliable and useful information during a building's life cycle to AECO stakeholders.

DT can be applied to any physical asset at any given time. For historical assets which have been completed many years or decades ago and do not yet have any digital records, DT can help to start and keep a record of their performance for better maintenance and renovation. For facilities under construction, a dynamic DT can support real-time progress monitoring, quality control, diagnostics, and prognostics. In addition, DT can also be used in the future for capital investment projects before the design and construction of the facility, as it provides an efficient way to simulate the performance of a building and aid the decision-making process.

The way the physical and the digital twins are synchronised in real use cases depends on the purpose of the DT, which also determines the content of DTs (i.e., the elements and processes to be digitised, the level of detail required, how frequent the model is supposed to be updated, etc.). As the concept of digital twins is broad, it is impractical to propose a precise and detailed definition of a digital twin that covers everything without considering its purpose. Some potential applications of DTs relevant to deep building renovations are presented hereafter.

Example 1: Condition Monitoring
A DT can be used to monitor the current condition of a building. By capturing geometric information through different sensors, the current condition of the asset can be visualised and represented by the DT. The geometry of facilities can be monitored by comparing the current condition with previous asset conditions over time, which allows a DT to give maintenance suggestions to the asset holders and managers (Hu et al., 2023).

Apart from monitoring the geometry change of a discrete asset, DT can also be used to monitor more complex large-scale systems, for example, the sewer system of a city. In this context, predictive maintenance operations can be utilised to identify potential blockages. Similarly, the current state of flow in pipes can be recorded and compared with the historical values to predict or locate disruptions in the system. Predictive maintenance recommendations or alerts can be sent to facility managers for more informed and timely decision-making.

Example 2: Facility Management
There is a very broad spectrum of facility operation, which includes but is not limited to operation management of mechanical, electrical, and plumbing (MEP) components in a facility (Z. Hu et al., 2016; Cheng et al., 2020), internal environment monitoring (Cao et al., 2015), and working productivity (Meerman et al., 2014). With the increasing adoption of the Internet of Things (IoT) and artificial intelligence (AI) which are key components supporting DTs, facility management is becoming more and more intelligent. Similarly, augmented reality (AR) and virtual reality (VR) can be used in conjunction with DT to visualise the built environment and improve efficiency (Baek et al., 2019; Chen et al., 2020; Chen et al., 2021; Zhang et al., 2020).

The concept of the digital twin is capable of embedding all these use cases in facility management according to the concept illustrated in Fig. 6.2. Relevant objects and values are captured and represented in a detailed digital model through capturing devices like laser scanners and cameras. By applying various IoT sensors such as thermometers, hygrometers, and carbon dioxide sensors, different values (like temperature, humidity, and carbon dioxide level) that represent internal environment conditions can be recorded and then updated in the digital model regularly. AI-relevant technologies can be used to help the process of creating the initial model as well as updating the model throughout a facility's life cycle. Facility managers can check the visually assistive information provided by AR and VR devices, which is able to lighten their workload and benefit working efficiency. From small-scale facilities, like offices, to large-scale urban environments, different sensors can be used to find how people exactly use these facilities and map occupant behaviour. With a better understanding of this data, the environmental conditions can be optimised, ultimately improving human wellness and living satisfaction.

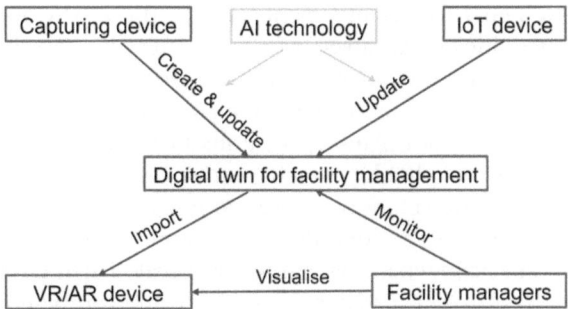

Fig. 6.2 Digital twin for facility management

Example 3: Environment Simulation
Digital twins can be used in the renovation phase of a project to simulate various scenarios without modifying the real asset. These scenarios can involve changing the natural light design, artificial lighting, heating simulation, and so forth. By only modifying facilities in the DT, the impact of these changes can be understood without implementing the modifications in the real world. VR/AR devices can make use of the DT to visualise the proposed designs and show the impact of changes and modifications (e.g., lighting). This improves the decision-making of renovation and enhances the communication between designers and clients. For instance, different lighting atmospheres can be visualised, helping designers to aesthetically assess the design and present the outcomes of the setup to their clients (Natephra et al., 2017).

6.5   Challenges to Digital Twin Adoption

Despite the fact that a DT is considered to offer benefits to all stakeholders of the built environment, some challenges hinder its adoption in real projects. Firstly, the effort involved in creating a digital twin is demanding, which undermines its feasibility and benefits. Many researchers are working on automating the process of digital twinning in the built environment in order to reduce human effort. The effort in the existing literature has been concentrated on reconstructing relatively large structural elements like ceilings, floors, and walls. MEP elements (such as fire alarms, emergency switches, etc.) should also be included in a DT, as these are essential elements for facility managers. In the repair and maintenance (R & M) activities of an asset, MEP costs usually constitute the largest share of the total cost (Adán et al., 2018). Therefore, a DT would be more valuable if it were to contain those elements. In addition, facility management also involves floor plans, space utilisation, asset location, and technical plants (D'Urso, 2011), which requires more accurate capture and modelling. Text information such as room numbers and serial numbers (IDs) of objects that can identify the corresponding asset instance is also beneficial, especially when managing large-scale facilities. These IDs represent the exact object instances in a facility and can be used to make the links between physical assets and DT much clearer. Currently, such activities are mainly performed manually in real projects. Some studies (e.g., Pan et al., 2022) in this area have started to emerge.

6.6   Conclusion and Future Direction

This chapter provides an overview of the background, definitions, generation, and applications of DT in the built environment generally and building renovation specifically. The state-of-the-art methods to create and update the geometry of digital twins were described. The potential applications of DTs, along with their advantages and current challenges, have been discussed with examples. The overarching conclusion is that DTs provide benefits and offer applications across the whole life cycle of built assets. Much research is still required to support the generation and the update of DTs, which is necessary to support the identified applications and unlock their respective benefits.

In the built environment, how to generate and update DTs precisely and efficiently to bring the benefits into real applications throughout the whole life cycle of a facility is still under research.

References

Adán, A., Quintana, B., Prieto, S. A., & Bosché, F. (2018). Scan-to-BIM for 'secondary' building components. *Advanced Engineering Informatics, 37*, 119–138. https://doi.org/10.1016/J.AEI.2018.05.001

Agapaki, E., & Brilakis, I. (2020). CLOI-NET: Class segmentation of industrial facilities' point cloud datasets. *Advanced Engineering Informatics, 45*(November 2019). https://doi.org/10.1016/j.aei.2020.101121

Agapaki, E., & Brilakis, I. (2021). Instance segmentation of industrial point cloud data. *Journal of Computing in Civil Engineering, 35*(6). https://doi.org/10.1061/(asce)cp.1943-5487.0000972

Agapaki, E., Miatt, G., & Brilakis, I. (2018). Prioritizing object types for modelling existing industrial facilities. *Automation in Construction, 96*(September), 211–223. https://doi.org/10.1016/j.autcon.2018.09.011

Baek, F., Ha, I., & Kim, H. (2019). Augmented reality system for facility management using image-based indoor localization. *Automation in Construction, 99*(November 2018), 18–26. https://doi.org/10.1016/j.autcon.2018.11.034

Cao, Y., Song, X., & Wang, T. (2015). Development of an energy-aware intelligent facility management system for campus facilities. *Procedia Engineering, 118*, 449–456. https://doi.org/10.1016/j.proeng.2015.08.446

Chen, H., Hou, L., Zhang, G., & Moon, S. (2021). Development of BIM, IoT and AR/VR technologies for fire safety and upskilling. *Automation in Construction, 125*(September 2020), 103631. https://doi.org/10.1016/j.autcon.2021.103631

Chen, K., Yang, J., Cheng, J. C. P., Chen, W., & Li, C. T. (2020). Transfer learning enhanced AR spatial registration for facility maintenance management. *Automation in Construction, 113*(July 2019), 103135. https://doi.org/10.1016/j.autcon.2020.103135

Cheng, J. C. P., Chen, W., Chen, K., & Wang, Q. (2020). Data-driven predictive maintenance planning framework for MEP components based on BIM and IoT using machine learning algorithms. *Automation in Construction, 112*(December 2019), 103087. https://doi.org/10.1016/j.autcon.2020.103087

D'Urso, C. (2011). Information integration for facility management. *IT Professional, 13*(6), 48–53. https://doi.org/10.1109/MITP.2011.100

El Saddik, A. (2018). Digital twins: The convergence of multimedia technologies. *IEEE Multimedia, 25*(2), 87–92. https://doi.org/10.1109/MMUL.2018.023121167

Hu, Z., Brilakis, I., Karlinsky,L., Michaeli, T., & Nishino, K. (2023). Computer Vision – ECCV 2022 Workshops Tel Aviv Israel October 23–27 2022 Proceedings Part VII PriSeg: IFC-Supported Primitive Instance Geometry Segmentation with Unsupervised Clustering Springer Nature Switzerland Cham 196–211. https://doi.org/10.1007/978-3-031-25082-8_13.

Hu, Z., Fathy, Y., & Brilakis, I. (2022). Geometry updating for digital twins of buildings: A review to derive a new geometry-based object class hierarchy. *Proceedings of the 2022 European Conference on Computing in Construction.* https://doi.org/10.35490/ec3.2022.155

Hu, Z. Z., Zhang, J. P., Yu, F. Q., Tian, P. L., & Xiang, X. S. (2016). Construction and facility management of large MEP projects using a multi-scale building information model. *Advances in Engineering Software, 100*, 215–230. https://doi.org/10.1016/j.advengsoft.2016.07.006

Lu, R., Brilakis, I., & Middleton, C. R. (2019). Detection of structural components in point clouds of existing RC bridges. *Computer-Aided Civil and Infrastructure Engineering*, *34*(3), 191–212. https://doi.org/10.1111/mice.12407

Maturana, D., & Scherer, S. (2015). VoxNet: A 3D Convolutional Neural Network for real-time object recognition. In *IEEE International Conference on Intelligent Robots and Systems*, *2015-Decem* (pp. 922–928). IEEE. https://doi.org/10.1109/IROS.2015.7353481

Meerman, A., Lellek, V., & Serbin, D. (2014). The path to excellence: Integrating customer satisfaction in productivity measurement in Facility Management. *International Journal of Facilities Management*, 201–211.

Monszpart, A., Mellado, N., Brostow, G. J., & Mitra, N. J. (2015). RAPter: Rebuilding man-made scenes with regular arrangements of planes. *ACM Transactions on Graphics*, *34*(4). https://doi.org/10.1145/2766995

Natephra, W., Motamedi, A., Fukuda, T., & Yabuki, N. (2017). Integrating building information modeling and virtual reality development engines for building indoor lighting design. *Visualization in Engineering*, *5*(1). https://doi.org/10.1186/s40327-017-0058-x

Ochmann, S., Vock, R., Wessel, R., & Klein, R. (2016). Automatic reconstruction of parametric building models from indoor point clouds. *Computers and Graphics (Pergamon)*, *54*, 94–103. https://doi.org/10.1016/j.cag.2015.07.008

Oesau, S., Lafarge, F., & Alliez, P. (2014). Indoor scene reconstruction using feature sensitive primitive extraction and graph-cut. *ISPRS Journal of Photogrammetry and Remote Sensing*, *90*, 68–82. https://doi.org/10.1016/j.isprsjprs.2014.02.004

Pan, Y., Braun, A., Borrmann, A., & Brilakis, I. (2021). Void-growing: A novel Scan-to-BIM method for Manhattan world buildings from point cloud. *Proceedings of the 2021 European Conference on Computing in Construction*, *2*(2018), 312–321. https://doi.org/10.35490/ec3.2021.162

Pan, Y., Braun, A., Brilakis, I., & Borrmann, A. (2022). Enriching geometric digital twins of buildings with small objects by fusing laser scanning and AI-based image recognition. *Automation in Construction*, *140*(February), 104375. https://doi.org/10.1016/j.autcon.2022.104375

Perez-Perez, Y., Golparvar-Fard, M., & El-Rayes, K. (2021). Scan2BIM-NET: Deep learning method for segmentation of point clouds for Scan-to-BIM. *Journal of Construction Engineering and Management*, *147*(9). https://doi.org/10.1061/(asce)co.1943-7862.0002132

Qi, C., Yi, L., Su, H., & Guibas, L. (2017b). PointNet++: Deep hierarchical feature learning on. In *NIPS'17: Proceedings of the 31st International Conference on Neural Information Processing Systems*, *Dec* (pp. 5105–5114). ACM Digital Library.

Qi, C. R., Su, H., Mo, K., & Guibas, L. J. (2017a). PointNet: Deep learning on point sets for 3D classification and segmentation. In *Proceedings—30th IEEE Conference on Computer Vision and Pattern Recognition, CVPR 2017, 2017-Janua* (pp. 77–85). IEEE. https://doi.org/10.1109/CVPR.2017.16

Sacks, R., Brilakis, I., Pikas, E., Xie, H. S., & Girolami, M. (2020). Construction with digital twin information systems. *Data-Centric Engineering, 1.* https://doi.org/10.1017/dce.2020.16

Sanchez, V., & Zakhor, A. (2012). Planar 3D modeling of building interiors from point cloud data. In *Proceedings—International Conference on Image Processing, ICIP* (pp. 1777–1780). IEEE. https://doi.org/10.1109/ICIP.2012.6467225

Stambler, A., & Huber, D. (2015). Building modeling through enclosure reasoning. In *Proceedings—2014 International Conference on 3D Vision Workshops, 3DV 2014* (pp. 118–125). ACM Digital Library. https://doi.org/10.1109/3DV.2014.65

Tao, F., Zhang, H., Liu, A., & Nee, A. Y. C. (2019). Digital twin in industry: State-of-the-Art. *IEEE Transactions on Industrial Informatics, 15*(4), 2405–2415. https://doi.org/10.1109/TII.2018.2873186

Tran, H., Khoshelham, K., Kealy, A., & Díaz-Vilariño, L. (2019). Shape grammar approach to 3D modeling of indoor environments using point clouds. *Journal of Computing in Civil Engineering, 33*(1). https://doi.org/10.1061/(asce)cp.1943-5487.0000800

Xiao, J., & Furukawa, Y. (2014). Reconstructing the world's museums. *International Journal of Computer Vision, 110*(3), 243–258. https://doi.org/10.1007/s11263-014-0711-y

Zhang, Y., Liu, H., Kang, S. C., & Al-Hussein, M. (2020). Virtual reality applications for the built environment: Research trends and opportunities. *Automation in Construction, 118*(May), 103311. https://doi.org/10.1016/j.autcon.2020.103311

CHAPTER 7

Additive Manufacturing and the Construction Industry

Mehdi Chougan, Mazen J. Al-Kheetan,
and Seyed Hamidreza Ghaffar

Abstract Additive manufacturing (AM), including 3D printing, has the potential to transform the construction industry. AM allows the construction industry to use complex and innovative geometries to build an object, building block, wall, or frame from a computer model. As such, it has potential opportunities for the construction industry and specific applications in the deep renovation process. While AM can provide significant benefits in the deep renovation process, it is not without its own environmental footprint and barriers. In this chapter, AM is defined, and the main materials used within the construction industry are outlined. This chapter

M. Chougan • S. H. Ghaffar (✉)
Department of Civil and Environmental Engineering,
Brunel University London, London, UK
e-mail: mehdi.chougan2@brunel.ac.uk; seyed.ghaffar@brunel.ac.uk

M. J. Al-Kheetan
Department of Civil and Environmental Engineering, Mutah University,
Mu'tah, Jordan
e-mail: mazen.al-kheetan@mutah.edu.jo

© The Author(s) 2023   97
T. Lynn et al. (eds.), *Disrupting Buildings*, Palgrave Studies in
Digital Business & Enabling Technologies,
https://doi.org/10.1007/978-3-031-32309-6_7

also explores the benefits and challenges of implementing AM within the construction industry before concluding with a discussion of the future areas of development for AM in construction.

Keywords Additive manufacturing • 3D printing technology • Construction industry • 3D concrete printing

7.1   INTRODUCTION

Additive manufacturing (AM) is the process of fabricating three-dimensional (3D) physical objects by connecting materials together in a layer-based manner following a specific computer design (Guo & Leu, 2013). The concept of AM was first introduced by Chuck Hull (1984), who used ultraviolet (UV) light to harden a layer of a liquid polymer (Wong & Hernandez, 2012). In recent years, AM has evolved to include a wide range of solutions and techniques, including selective laser sintering (SLS), direct metal laser sintering (DMLS), laser engineered net shaping (LENS), electron beam melting (EBM), fused deposition modelling (FDM), and digital light processing (DLP) (Albar et al., 2020). These methods enable the use of different materials in AM such as metals, composites, ceramics, and polymers and the production of end-parts that are capable of serving different purposes (Albar et al., 2020). The rapid development of AM has encouraged researchers and practitioners to adopt this technology in the construction sector as a cost-effective solution to create various structural components, regardless of their complexity, with minimum waste (Lyu et al., 2021).

In the construction sector, a significant focus of research and development was observed towards the development of different AM methods to cope with the unique characteristics of cementitious materials. These mostly include material extrusion and particle-bed processes as well as other generative approaches such as Smart Dynamic Casting (Paolini et al., 2019). Aggregate-based materials such as concrete are most commonly used in AM for the construction industry (Paolini et al., 2019). According to recent estimates, the value of the AM market for concrete printing was over $310 million in 2019 and is expected to reach $40 billion by 2027 with an annual growth rate of 116% (Pawar & Rohit Sawant, 2020). These figures suggest that AM will be rapidly and globally adopted by the construction sector, driven by the promise of reduced environmental impact, support for more complex designs, and more cost-effective construction (Mart et al., 2022). It is important to note that while AM processes are less labour-intensive, the adoption of AM in construction is expected to result in significant job creation, including new high-value roles, for example, 3D printer manufacturing and maintenance engineers, mixture designers, materials suppliers, and specialist software developers (Avrutis et al., 2019).

The remainder of this chapter introduces and defines AM and provides an overview of the main benefits of AM as well as its main applications in construction and deep renovation[1] projects. Finally, practical challenges in the implementation of additive manufacturing are summarised, and upcoming advancements are briefly discussed in the final section.

7.2   ADDITIVE MANUFACTURING IN CONSTRUCTION AND DEEP RENOVATION

Significant advancements have been made in concrete 3D printing in recent years thanks to the introduction of a variety of different materials in producing concrete mixtures. Ordinary Portland cement (OPC) was the first material adopted by AM to produce full-scale printed concrete structures (Chougan et al., 2021). There are, however, concerns regarding the impact of OPC on the environment, which remains an issue with its implementation in AM. Cement production accounts for 5–7% of the total world CO_2 emissions (Chougan et al., 2021). In order to achieve a sustainable built environment and reduce CO_2 emissions, many researchers have suggested the implementation of alkali-activated materials (AAMs) as they can entirely replace OPC and produce a low-carbon binder (Chougan et al., 2021). In this case, materials such as metakaolin are used as aluminosilicate cementitious binders along with activators such as potassium silicate, sodium metasilicate, and potassium hydroxide to obtain AAM binders capable of building successful 3D printed structures (Alghamdi et al., 2019). Others have suggested enhancing AAMs' rheological properties by integrating modifying agents and additives in the mixtures like polypropylene (PP), polyvinyl alcohol (PVA), nano-graphite (NG), halloysite clay minerals, and attapulgite to improve the buildability, printability, and mechanical performance of AAMs for 3D printing (Chougan et al., 2021; Chougan et al., 2022).

The application of AM is not limited to the 3D printing of cementitious composites. Aside from cementitious composites, other categories of materials, such as polymers and metals, have also been used, particularly in renovation works. With the continuous development of AM technology,

---

[1] Chapter 1 in this book provides a detailed definition of deep renovation.

the customisation of parts and components needed for particular purposes in renovation projects became possible. For instance, the production and installation of precast concrete façade sections can be particularly challenging due to their complexity and the wide variation in their configurations in different buildings. In this context, AM could enhance the quality of the produced façade sections due to its higher degree of flexibility compared to standard production methods while also minimising post-installation problems such as air and water leakages. AM can also be used to print moulds that have the ability to produce façade sections with efficient passive shading (Harris, 2022).

AM is being scaled increasingly. Big area additive manufacturing (BAAM), a 3D-printing process similar to FDM, has been developed to construct segments of cylindrical single-floored building components out of polymer materials such as neat ABS and CF-ABS (Biswas et al., 2017). In addition, robotic 3D metal printing, also known as wire arc additive manufacturing (WAAM), can be used to fabricate highly tailored and engineered steel connectors for large structures in the construction sector (Xin et al., 2021).

More examples of the cementitious composites, polymer, and metal additive manufacturing technologies in building structures can be found in Table 7.1.

7.3   BENEFITS OF ADDITIVE MANUFACTURING IN CONSTRUCTION

Historically, the construction industry was characterised by high energy consumption (i.e., 40% of the global energy consumption), high solid waste production (i.e., 40% of the global waste production), high greenhouse gases emission (i.e., 38% of the global CO_2 emission), and high water depletion (i.e., 12% of the global water depletion) (Comstock et al., 2012). It has an undeniably high environmental footprint. Growing public interest in sustainability highlights the necessity for novel construction techniques and materials to mitigate traditional construction's high environmental impacts. AM technology represents one possible way for construction companies to use available resources more efficiently. In fact, one of the main advantages of AM is the minimisation of raw materials consumption, which reduces the level of waste generated during construction (Yao et al., 2020; Valente et al., 2022).

A second related advantage of AM compared to traditional construction methods is the capacity to produce complicated large-scale structures

Table 7.1 Various materials and technologies used in additive manufacturing

Material category	Technology	AM process	Reference
Cementitious composites	OPC-based 3D printing	Extrusion-based	Cuevas et al. (2021)
	AAM 3D printing	Extrusion-based	Chougan et al. (2020)
Polymers	Qingdao Unique Products Develop	Extrusion-based	Feng and Yuhong (2014)
	BAAM	Extrusion-based	Love (2015)
	C-Fab	Extrusion-based	Technology (2017)
	Digital Construction Platform (DCP)	Extrusion-based	Keating et al. (2014)
	FreeFAB™ Wax	Extrusion-based	Gardiner et al. (2016)
Metals	Maraging steel	Powder bed fusion	Galjaard et al. (2015)
	Multiple	Powder bed fusion	Mrazovic (2016)
	Stainless steel	Direct energy deposition	Joosten (2015)
	Aluminium	Powder bed fusion	Strauss and Knaack (2015)

while also minimising raw materials waste by lowering or eliminating the necessity of conventional formworks (Wangler et al., 2016). The increasing use of cementitious materials (e.g., concrete) in construction, along with the high costs of formwork production, emphasises the value of additive manufacturing technologies in constructing complex structures. Furthermore, the ability to fabricate complex objects enables building structures to possess "multi-functionality" by facilitating the integration of services, including piping, insulation, and electrical setups, and offering a secondary function through its complex geometry, such as instinct thermal insulation (De Schutter et al., 2018). This may be particularly beneficial in the context of deep renovation where the number of building elements to be replaced is quite large and existing building constraints make the installation of different individual elements quite challenging. As the structure becomes more complex, AM technology becomes more advantageous. In the same way, AM may be less cost-effective and less environmentally beneficial for more "standard" designs (Labonnote & Rüther, 2017).

Finally, as AM processes remove the need for conventional energy-consuming processes and labour-intensive activities like concrete pumping and casting, shuttering, material logistics, and steel fixing, it reduces the costs of on-site assembly and construction, minimises human error, and improves productivity (Avrutis et al., 2019).

7.4   PRACTICAL CHALLENGES FOR AM IN THE CONSTRUCTION INDUSTRY

Despite the benefits of AM to the construction industry, there are a series of major challenges to its implementation, which could hinder adoption. Firstly, the high cost of obtaining 3D printing equipment, as well as printers' transportation and logistics, could arguably represent a significant obstacle to the widespread application of 3D printing technology in the construction industry. Despite the technological advantages, many construction companies are still unable to justify or afford an investment in 3D printing equipment.

Secondly, while AM technology reduces human errors and the need for workers on construction sites, finding qualified individuals to work with AM remains difficult (Deloitte, 2016). In addition, the shorter production and installation time comes at the cost of a longer design phase, which requires significantly higher effort and specialised modelling skills (Buswell et al., 2018). These labour supply problems are multifaceted. They are driven by a decline in the attractiveness of the manufacturing and construction sector, a lack of labour supply from the education sector with sufficient STEM skills and knowledge, a shortage of AM-specific training programmes, and a general lack of AM knowledge and culture in many construction and construction-related manufacturing companies (Deloitte, 2016). Where skilled labour does exist, firms may face significant upskilling, skills maintenance, and retention challenges until the AM skills and training gaps are addressed (Deloitte, 2016).

Thirdly, there is a general lack of regulation, standardisation, and testing of AM printing structures and materials. Standards in AM facilitate technology adoption, boost confidence in the quality and safety of AM processes, materials, and outputs, and support the competitiveness of AM and construction companies (Martínez-García et al., 2021). While standards have been developed by a wide range of organisations, for example, the German Society of Mechanical Engineers, the ISO, and the American

Society for Testing and Materials (ASTM), there would appear to be some challenges in aligning existing standards for testing the mechanical properties of more traditional materials and manufactured polymers and composites, and those generated through AM (Forster, 2015; Martínez-García et al., 2021). Indeed, it is fair to say that the flexibility AM introduces in terms of design and material use complicates testing and standards. Martínez-García et al. (2021) note that despite significant efforts by the ISO and ASTM, AM technology requires specific standards in all the stages of the product development, including design, materials, manufacturing, and final part.

Finally, and somewhat contradictory to the benefits presented in the previous section, the environmental impact of AM may not be entirely positive. AM is still at an early stage of development and use in the construction sector. Much AM use still involves the use of environmentally hazardous substances (e.g., cement) in considerable quantities as well as substantial equipment and non-eco-friendly manufacturing (Agustí-Juan & Habert, 2017; Agustí-Juan et al., 2017). AM units are often powered by lithium batteries and the electricity consumption throughout the fabrication process may offset the waste reduction and the other environmental benefits generated by AM (Agustí-Juan & Habert, 2017; Agustí-Juan et al., 2017).

7.5   Future Areas of Development

AM is particularly economically beneficial for large-scale building developments due to the enhanced geometrical freedom enabled by this technology. Compared to traditional construction methods, AM technology provides architecture designers the geometric freedom to create ideal complex structures while minimising the use of materials (Labonnote et al., 2016). However, while AM construction methods have been extensively adopted in real applications, there is still a lack of knowledge regarding large-scale AM. As a result, large-scale AM can be considered an escalated challenge compared to lab-scale 3D printing. Large-scale AM is typically more complicated than lab-scale 3D printing, as several practical construction challenges must be addressed. Large-scale AM involves a set of discrete technologies and thus requires consideration of a very different set of parameters, not least materials, reinforcing admixtures, economics, environmental optimisations, structural limitations, and 3D printing system design (Xiao et al., 2021). The majority of the existing studies concentrated on 3D printing of cementitious composites containing fine aggregate (i.e., mortar); however, cementitious composites with coarse aggregate (i.e., concrete) are attracting considerable interest because of their remarkable mechanical and cost-efficiency advantages (Xiao et al., 2021). Therefore, further investigation is required to determine the impact of using coarse aggregates to move towards cementitious concretes in order to fulfil the large-scale 3D printing requirements.

4D printing, a novel approach that includes a fourth dimension (i.e., time and smart behaviour), can allow 3D-printed items to transform their geometry and behaviour throughout time in response to specific conditions such as radiation, light, and temperature. The smart behaviour of 4D printing in shifting configurations for self-assembly, multi-functionality, and self-repair is a crucial breakthrough in AM technology. While 4D printing delivers all of the advantages of 3D printing, its use in the construction sector is in its infancy, posing obstacles such as a considerable need for improved computer analysis, new design concepts, structure validation, and standardisation (Pan & Zhang, 2021).

7.6   CONCLUSION

AM technology represents a valuable innovation in the construction sector and is gaining popularity. There are many benefits to AM, such as its potential to significantly reduce the consumption rate of raw materials, reduce the generated waste during construction, lower CO_2 emissions, reduce labour costs, minimise human errors, and improve productivity. Many complex designs, at a building or part-level, that previously were considered too problematic or costly for execution on-site can be easily implemented with the help of AM technology. Widespread adoption is not without challenges. In fact, some issues still exist in relation to process, materials, geometric complexity, software and building integration, and the standards associated with these elements. In order to capitalise on the impact of AM, additional research is needed to support the better integration of this technology in the construction sector. Moreover, to enable the rapid growth of this technology, standardised testing and quality control methods should be established to improve information sharing and benchmarking. Finally, without a pipeline of qualified labour, the full potential of AM will not be realised. This will be a key challenge to overcome if the technology is to be pushed further into full-scale industrialisation.

REFERENCES

Agustí-Juan, I., & Habert, G. (2017). Environmental design guidelines for digital fabrication. *Journal of Cleaner Production, 142*, 2780–2791. https://doi. org/10.1016/j.jclepro.2016.10.190

Agustí-Juan, I., Müller, F., Hack, N., Wangler, T., & Habert, G. (2017). Potential benefits of digital fabrication for complex structures: Environmental assessment of a robotically fabricated concrete wall. *Journal of Cleaner Production, 154*, 330–340. https://doi.org/10.1016/j.jclepro.2017.04.002

Albar, A., Chougan, M., Al-Kheetan, M. J., Swash, M. R., & Ghaffar, S. H. (2020). Effective extrusion-based 3D printing system design for cementitious-based materials. *Results in Engineering, 6*(April), 100135. https://doi.org/10.1016/j.rineng.2020.100135

Alghamdi, H., Nair, S. A. O., & Neithalath, N. (2019). Insights into material design, extrusion rheology, and properties of 3D- printable alkali-activated fly ash-based binders. *Materials & Design, 167*, 107634. https://doi.org/10.1016/j.matdes.2019.107634

Avrutis, D., Nazari, A., & Sanjayan, J. G. (2019). Industrial adoption of 3D concrete printing in the Australian market. In *3D concrete printing technology*. Elsevier Inc. https://doi.org/10.1016/b978-0-12-815481-6.00019-1

Biswas, K., Rose, J., Eikevik, L., Guerguis, M., Enquist, P., Lee, B., Love, L., Green, J., & Jackson, R. (2017). Additive manufacturing integrated energy-enabling innovative solutions for buildings of the future. *Journal of Solar Energy Engineering, Transactions of the ASME, 139*(1), 1–10. https://doi.org/10.1115/1.4034980

Buswell, R. A., De Silva, W. R. L., Jones, S. Z., & Dirrenberger, J. (2018). Cement and Concrete Research 3D printing using concrete extrusion: A roadmap for research. *Cement and Concrete Research, 112*(May), 37–49. https://doi.org/10.1016/j.cemconres.2018.05.006

Chougan, M., Ghaffar, S. H., Nematollahi, B., Sikora, P., Dorn, T., Stephan, D., Albar, A., & Al-Kheetan, M. J. (2022). Effect of natural and calcined halloysite clay minerals as low-cost additives on the performance of 3D-printed alkali-activated materials. *Materials and Design, 223*. https://doi.org/10.1016/j.matdes.2022.111183

Chougan, M., Ghaffar, S. H., Sikora, P., Chung, S. Y., Rucinska, T., Stephan, D., Albar, A., & Swash, M. R. (2021). Investigation of additive incorporation on rheological, microstructural and mechanical properties of 3D printable alkali-activated materials. *Materials and Design, 202*. https://doi.org/10.1016/j.matdes.2021.109574

Chougan, M., Hamidreza Ghaffar, S., Jahanzat, M., Albar, A., Mujaddedi, N., & Swash, R. (2020). The influence of nano-additives in strengthening mechanical performance of 3D printed multi-binder geopolymer composites. *Construction and Building Materials, 250*, 118928. https://doi.org/10.1016/j.conbuildmat.2020.118928

Comstock, M., Garrigan, S., Pouffary, T. D., & Feraudy, J. (2012). *Building design and construction: Forging resource efficiency and sustainable development.* https://www.unep.org/explore-topics/resource-efficiency/what-we-do/cities/sustainable-buildings

Cuevas, K., Chougan, M., Martin, F., Ghaffar, S. H., Stephan, D., & Sikora, P. (2021). 3D printable lightweight cementitious composites with incorporated waste glass aggregates and expanded microspheres—Rheological, thermal and mechanical properties. *Journal of Building Engineering, 44*(February). https://doi.org/10.1016/j.jobe.2021.102718

De Schutter, G., Lesage, K., Mechtcherine, V., Nerella, V. N., Habert, G., & Agusti-Juan, I. (2018). Vision of 3D printing with concrete—Technical, economic and environmental potentials. *Cement and Concrete Research, 112*(November 2017), 25–36. https://doi.org/10.1016/j.cemconres.2018.06.001

Deloitte. (2016). *3D opportunity for the talent gap—Additive manufacturing and the workforce of the future.* Deloitte University Press. https://www2.deloitte.com/content/dam/insights/us/articles/3d-printing-talent-gap-workforce-development/ER_3062-3D-opportunity-workforce_MASTER.pdf

Feng, L., & Yuhong, L. (2014). Study on the status quo and problems of 3D printed buildings in China. *Global Journal of Human-Social Science Research, 14*(5), 1–4.

Forster, A. M. (2015). *Materials testing standards for additive manufacturing of polymer materials: State of the art and standards applicability.* NIST.

Galjaard, S., Hofman, S., & Ren, S. (2015). New opportunities to optimize structural designs in metal by using additive manufacturing. In *Advances in architectural geometry 2014* (pp. 79–93). Springer. https://doi.org/10.1007/978-3-319-11418-7_6

Gardiner, J. B., Janssen, S., & Kirchner, N. (2016). A realisation of a construction scale robotic system for 3D printing of complex formwork. In *ISARC 2016—33rd International Symposium on Automation and Robotics in Construction (ISARC)* (pp. 515–521). The International Association for Automation and Robotics in Construction. https://doi.org/10.22260/isarc2016/0062

Guo, N., & Leu, M. C. (2013). Additive manufacturing: Technology, applications and research needs. *Frontiers of Mechanical Engineering, 8*(3), 215–243. https://doi.org/10.1007/s11465-013-0248-8

Harris, C. (2022). *Pathway to zero energy windows: Advancing technologies and market adoption.* April.

Hull, C. W. (1984). Apparatus for production of three-dimensional objects by stereolithography. United States Patent, Appl., No. 638905, Filed.

Joosten, S. K. (2015). *Printing a stainless steel bridge: An exploration of structural properties of stainless steel additive manufactures for civil engineering purposes.*

Structural   Engineering.   https://repository.tudelft.nl/islandora/object/
uuid:b4286867-9c1c-40c1-a738-cf28dd7b6de5?collection=education
Keating, S., Spielberg, N. A., Klein, J., & Oxman, N. (2014). A compound arm
approach to digital construction. In *Robotic fabrication in architecture, art and
design*. Springer International Publishing. https://doi.org/10.1007/
978-3-319-04663-1_7
Labonnote, N., & Rüther, P. (2017). Additive manufacturing: An opportunity for
functional and sustainable constructions. In *Challenges for technology innovation: An agenda for the future—Proceedings of the International Conference on
Sustainable Smart Manufacturing, S2M 2016, September* (pp. 201–206). Taylor
& Francis. https://doi.org/10.1201/9781315198101-41
Labonnote, N., Rønnquist, A., Manum, B., & Rüther, P. (2016). Additive con-
struction: State-of-the-art, challenges and opportunities. *Automation in
Construction, 72*, 347–366. https://doi.org/10.1016/j.autcon.2016.08.026
Love, L. J. (2015). *Utility of Big Area Additive Manufacturing (BAAM) for the
rapid manufacture of customized electric vehicles*. Oak Ridge National Laboratory
(ORNL), Manufacturing Demonstration Facility (MDF). https://doi.org/
10.2172/1209199
Lyu, F., Zhao, D., Hou, X., Sun, L., & Zhang, Q. (2021). Overview of the development of 3D-printing concrete: A review. *Applied Sciences, 11*(21), 9822.
https://doi.org/10.3390/app11219822
Mart, A., Garc, R., Muñoz-sanguinetti, C., Felipe, L., & Auat-cheein, F. (2022).
Recent developments and challenges of 3D-printed construction: A review of
research fronts. *Buildings, 12*(2), 229. https://doi.org/10.3390/buildings
12020229
Martínez-García, A., Monzón, M., & Paz, R. (2021). Standards for additive manufacturing technologies: Structure and impact. In *Additive manufacturing*
(pp. 395–408). Elsevier.
Mrazovic, N. (2016). *Feasibility study to 3D print a full scale curtain wall frame as
a single element*. http://cife.stanford.edu/sites/default/files/FeasibilityStudy
3DPrinting4Permasteelisa.pdf
Pan, Y., & Zhang, L. (2021). Roles of artificial intelligence in construction engineering and management: A critical review and future trends. *Automation in
Construction, 122*(November 2020), 103517. https://doi.org/10.1016/j.
autcon.2020.103517
Paolini, A., Kollmannsberger, S., & Rank, E. (2019). Additive manufacturing in
construction: A review on processes, applications, and digital planning methods. *Additive Manufacturing, 30*(July), 100894. https://doi.org/10.1016/j.
addma.2019.100894
Pawar, D., & Rohit Sawant, O. S. (2020). *3D concrete printing market by printing
type (gantry system and robotic arm), technique (extrusion-based and powder-
based), and end-use sector (residential, commercial, and infrastructure): Global
opportunity analysis and industry forecast, 2020–2027*. https://www.alliedmar-
ketresearch.com/3d-concrete-printing-market

Strauss, H., & Knaack, U. (2015). Additive manufacturing for future facades: The potential of 3D printed parts for the building envelope. *Journal of Facade Design and Engineering, 3*(3–4), 225–235.

Technology, B. (2017). *Cellular fabrication.* http://www.branch.technology

Valente, M., Sambucci, M., Chougan, M., & Ghaffar, S. H. (2022). Reducing the emission of climate-altering substances in cementitious materials: A comparison between alkali-activated materials and Portland cement-based composites incorporating recycled tire rubber. *Journal of Cleaner Production, 333*(November 2021), 130013. https://doi.org/10.1016/j.jclepro.2021. 130013

Wangler, T., Lloret, E., Reiter, L., Hack, N., Gramazio, F., Kohler, M., Bernhard, M., Dillenburger, B., Buchli, J., Roussel, N., & Flatt, R. (2016). Digital concrete: Opportunities and challenges. *RILEM Technical Letters, 1*(November), 67. https://doi.org/10.21809/rilemtechlett.2016.16

Wong, K. V., & Hernandez, A. (2012). A review of additive manufacturing. *ISRN Mechanical Engineering, 2012,* 1–10. https://doi.org/10.5402/2012/ 208760

Xiao, J., Ji, G., Zhang, Y., Ma, G., Mechtcherine, V., Pan, J., Wang, L., Ding, T., Duan, Z., & Du, S. (2021). Large-scale 3D printing concrete technology: Current status and future opportunities. *Cement and Concrete Composites, 122*(December 2020), 104115. https://doi.org/10.1016/j.cemconcomp. 2021.104115

Xin, H., Tarus, I., Cheng, L., Veljkovic, M., Persem, N., & Lorich, L. (2021, December). Experiments and numerical simulation of wire and arc additive manufactured steel materials. In *Structures* (vol. 34, pp. 1393–1402). Elsevier.

Yao, Y., Hu, M., Di Maio, F., & Cucurachi, S. (2020). Life cycle assessment of 3D printing geo-polymer concrete: An ex-ante study. *Journal of Industrial Ecology, 24*(1), 116–127. https://doi.org/10.1111/jiec.12930

CHAPTER 8

Intelligent Construction Equipment and Robotics

Alessandro Pracucci, Laura Vandi,
and SeyedReza RazaviAlavi

Abstract With recent advancement in software, hardware, and computing technologies, applications of intelligent equipment and robots (IER) are growing in the construction industry. This chapter aims to review key advantages, use cases and barriers of adopting IER in construction and renovation projects. The chapter evaluates the maturity of available IER technologies in the market and discusses the key concerns and barriers for adopting IER such as the unstructured and dynamic nature of construction sites limiting mobility and communication of IER, hazards of human-robot interactions, training and skills required for operating and collaborating with IER, and cybersecurity concerns. Finally, the chapter

A. Pracucci • L. Vandi (✉)
Focchi Spa Unipersonale, Poggio Torriana, Italy
e-mail: a.pracucci@focchi.it; l.vandi@focchi.it

S. RazaviAlavi
Department of Mechanical and Construction Engineering,
Northumbria University, Newcastle upon Tyne, UK
e-mail: reza.alavi@northumbria.ac.uk

© The Author(s) 2023
T. Lynn et al. (eds.), *Disrupting Buildings*, Palgrave Studies in
Digital Business & Enabling Technologies,
https://doi.org/10.1007/978-3-031-32309-6_8

proposes a framework for implementing IER that helps in their benefits by defining relevant metrics while considering their pitfalls in terms of quality, safety, time, and cost. This framework assists practitioners in decision-making for adopting IER in their construction operation.

Keywords Robotics • Construction • Safety • Monitoring • Quality control • Assessment framework

8.1   Key Definitions and Concepts

Table 8.1 provides a summary of key definitions and concepts related to the use of intelligent construction equipment and robotics in the construction industry.

Table 8.1   Key definitions and concepts

Construction Automation and Robotics (CAR)	A field of research and development focused on automating construction processes; construction automation deals with applying the principles of industrial automation to the construction sector (Saidi et al., 2016).
Single Task Construction Robots	Robots or automated devices that are developed primarily for performing a specific task on the construction site (Hu et al., 2020).
Integrated Robotised Construction Sites	Construction sites in which multiple robots/machines collaborate to build an entire structure (Saidi et al., 2016).
Teleoperation	A robot technology where a human operator controls a remote robot (Lichiardopol, 2007).
Programmable Construction Machines	A type of machine of which operator can change the activities to be accomplished within certain constraints either by selecting from a preprogrammed menu of functions or by teaching the machine a new function (Saidi et al., 2016).
Intelligent Systems	Software programs that syndicate the knowledge of experts and attempt to resolve distinct problems by imitating the reasoning processes of experts (Irani & Kamal, 2014).
Cobots	A system that amplifies or assists human skills, while performing tasks that require both the capacity of a human and the accuracy of a robot (Melo et al., 2012).
Exoskeletons	Emerging wearable technologies involved in the entire construction sector phase which allow to facilitate construction workers to lift heavy weights by reducing fatigue and site injuries and improving the work productivity (Kim et al., 2019).

(*continued*)

Table 8.1 (continued)

Robots	Devices that execute specific operations either autonomously or under an operator's direct control. The use of robots on construction sites is still very limited but the robotics production market is predicted to grow steadily over the next few years (Davila Delgado et al., 2019; European Construction Sector Observatory, 2021).
Unmanned Aerial Vehicle (UAV)	Commonly known as "drones," programmed technologies which can perform air operations reaching dangerous places for humans. Their use results in evident economic savings and environmental benefits while reducing the risk to human life (Outay et al., 2020).

8.2   INTRODUCTION

The construction industry plays a crucial role in ensuring job creation, driving economic growth, and providing solutions to address environmental, social, and economic challenges. The market value of the construction sector represents between 9% and 15% of GDP in most countries (Davila Delgado et al., 2019). Despite its huge economic importance, the construction industry is traditionally slow to change and consequently beset with inefficiencies resulting in lower productivity levels compared to other sectors (Davila Delgado et al., 2019). However, despite the complexity and fragmentation of the construction industry and the difficulties of coordinating the wide numbers of players and their tasks that slow down the introduction of innovative solutions, the construction sector has evolved in the last 25 years. This is especially driven by digital technologies and automation providing the construction industry with an opportunity to find innovative solutions to some of its rooted challenges. These innovations spanned across the whole project lifecycle, from design and engineering, through manufacturing and construction, to operation and maintenance, and retrofit/reuse/end-of-life. Among these, robotics is an emerging technological branch that can have an impact in construction areas such as off-site production, installation activities on-site, and operation and maintenance. This chapter will provide key insights about the digital transformation enabled by IER solutions in construction sites, analyze their current applications, limitations, and future developments, and propose an assessment framework to support construction actors in the decision-making process into the gradual adoption of IER for performing specific tasks.

8.3   Advantages and Benefits of IER

8.3.1   Improving Safety

The incident rate in the construction industry is the highest among various major industries in many countries (Choi et al., 2011). In the US, 25% of the fatal work injuries in 2020 belong to the construction sector (U.S. Bureau of Labor, 2021). In Great Britain, 1.8% of the construction workers reported a musculoskeletal disorder, which is the highest rate among the industries with similar work activities (Health and Safety Executive, 2021). Replacing humans by semi-autonomous and autonomous robots for undertaking unsafe tasks can reduce the number of incidents (Ilyas et al., 2021). Robots can be used for automating unsafe activities including heavy lifting and on-site inspection in dangerous work environments such as underground mines (Zimroz et al., 2019) and bridges (Lin et al., 2021). To reduce musculoskeletal injuries and physical fatigue of construction workers caused by repetitive and prolonged manual tasks, exoskeleton is being used for augmenting workers' physical ability (Brissi et al., 2022). Safety inspections and monitoring are other tasks that can be automated by robots for detecting unsafe locations (Martinez et al., 2020) and Personal Protective Equipment (PPE) on job sites (Ilyas et al., 2021).

8.3.2   Improving Productivity

Productivity growth has been a major concern in the construction industry as it was only one-third of the average total economy productivity growth over the past 20 years (Ribeirinho et al., 2020). Productivity of the construction industry can be improved by automating and robotising repetitive and labour-intensive activities. Autonomous transportation of construction materials by robots can improve productivity and eliminate human errors in these processes (Chea et al., 2020). For heavy lifting, robotic crane systems could improve productivity by 9.9–50% (Lee et al., 2009). The examples of IER applications for automation of different construction activity types are presented in Table 8.2.

8.3.3   Addressing Skilled Worker Shortage

Skilled worker shortage has been one of main issues in the construction industry over the past few years (Kim et al., 2020). The growing demand of construction workers and the aging workforces in many countries such as the UK (CITB, 2021; Green, 2021) are the main contributors to the

Table 8.2   IER application for improving productivity of different types of construction activities

Construction operation	Robot application
Masonry work	IER are used for automating bricklaying in masonry work. Hadrian X is the first mobile robotic bricklaying machine that uses 3D CAD model for accurately building masonry structures (FBR, 2022).
Precast concrete	IER are used for undertaking various tasks such as placing molds, reinforcement and distribution of concrete, and transportation of concrete formwork (Reichenbach & Kromoser, 2019; Saidi et al., 2016).
Steel component fabrication	IER are used for welding (Heimig et al., 2020), laser cutting (Bogue, 2008), bolting (Chu et al., 2013), and assembly (C. J. Liang et al., 2017) of steel components.
Timber construction	IER are used for cutting and drilling timber, and grasping, manipulating, and positioning building components (Eversmann et al., 2017; Willmann et al., 2016).

skilled worker shortage. In the long term, leveraging construction automation and replacing humans with IER can address this issue (Melenbrink et al., 2020). In addition, use of IER can address the challenges of the high labour wage in construction projects particularly in the metropolitan areas (Pan et al., 2020).

8.4   KEY USE CASES FOR INTELLIGENT CONSTRUCTION EQUIPMENT AND ROBOTICS

Although the impact of IER has not yet been fully realised in the construction industry (Carra et al., 2018), their applications are emerging to enhance construction productivity, safety management, quality control, and site planning issues. The first examples of construction robots were seen in the Japanese construction industry in the late 1970s and 1980s to supplement and replace workforce (Yilmaz & Metin, 2020). Construction automation and robotics application are classified in this chapter according to:

- Construction phase involvement—whether they are applied at the construction site (related to on-site activities) or at a factory for prefabrication activities (related to off-site activities) (Saidi et al., 2016). (Table 8.3);
- Level of autonomy—the second classification is based on the level of autonomy that IER technologies allow to perform (Table 8.4).

Table 8.3   Description of construction phase for IER classification

Off-site application	*Off-site* construction is widely used since the adoption of prefabrication approaches increase the control and the quality of the technological component manufactured. Indeed, the activities are conducted in a controlled environment as a factory with the consequence of reducing the risk of low quality during on-site installation. The adoption of IER solutions in a factory moves construction toward an industrialised sector with well-consolidated off-site activities.
On-site application	*On-site* execution is still a manual activity in many cases with the consequences of leading to problems such as unpredictable tasks and low levels of accuracy (Davila Delgado et al., 2019). The tasks during on-site stage are focused on the correct product installation, keep control of tasks advancement and monitoring with inspections activities the quality results. The traditional on-site activities require an appropriate level of labour skills to achieve the necessary efficiency in terms of construction duration and cost, and building quality (Yilmaz & Metin, 2020). On-site applications include:

- **Construction**—phase which involves the installation of different materials and construction actions (bricks laying, concrete formwork, timber frame as described in Table 8.2).
- **Inspection**—the objective of this task is to monitor the construction site activities in terms of time, quality, and cost. Technologies involved in this phase are equipped by a camera with the objectives to take pictures and share information regarding the construction site. Therefore, through the optimisation of the route, it reports in a regular range of time the work status verifying the correctness of installation.
- **Maintenance**—this stage includes the set of actions to preserve the integrity and the functionalities of the building during its life. The different technologies installed in the building or infrastructure, mainly the active ones with a higher degree of deterioration and the ones subjected to external interference (e.g., weather conditions, users' utilisation) that require a scheduled plan of maintenance and control of performance over time (Fig. 8.1).

Table 8.4   Description of autonomy level for IER classification

Teleoperated systems	The system includes remote and human control systems. This fulfills industrial situations where there is danger to the operator and where remote-controlled machinery is necessary.
Programmable Construction Machines (PCM)	It includes most construction equipment that is outfitted with sensors and mechanisms to augment operation by an onboard human operator.
Intelligent systems	It relates to unmanned construction robots which operate in either a semi- or a fully autonomous mode. This category also referred to the concept of adaptive manipulation, imitation learning, improvisatory control, and full autonomy.

Fig. 8.1 Allianz Tower while human workers are cleaning the façade. (Credit: Piermario Ruggeri-Focchi façade)

IER technologies can be further classified based on their technology readiness level (TRL) which identifies the maturity of the technologies within the market. In particular:

- TRL < 5—implies technologies which have been prototyped;
- TRL 6–7—implies technologies which have been tested and validated in an operational environment;
- TRL > 8—implies technologies which are widely used on market, indeed are considered actual system/process completed, and qualified through test and demonstration (pre-commercial demonstration).

Table 8.5 shows TRL for different IER technologies. The TRL level has been assigned based on market and academic research.

Table 8.5 TRL of the available IER technologies (TRL <5 listed as "✓", TRL 6–7 listed as "✓✓", TRL > 8 listed as "✓✓✓")

Classification of construction automation and robotics

Classification	Off-site activities	TRL	On-site activities					
			Construction	TRL	Inspection	TRL	Maintenance	TRL
Teleoperated systems	Exoskeleton	✓✓	Exoskeleton	✓✓	Drone	✓✓	Exoskeleton	✓
	–		Robotics arm	✓	Automatics monitoring	✓✓✓	Drone	✓
	–		Vehicles	✓	–		Automatics monitoring	✓✓✓
Programmable construction machines	Additive manufacturing	✓✓✓	Additive manufacturing	✓✓	Automatics monitoring	✓✓	Automatics monitoring	✓
	Robotics arm	✓	Robotics arm	✓✓	–		–	
	–		Integrated solution	✓	–		–	
Intelligent system	Robotics arm	✓✓	Robotics arm	✓✓	Automatics monitoring	✓	–	

The next subsections present some key examples of IER applications in the construction industry to highlight their significant impacts on various aspects of construction projects.

Additive manufacturing for construction phase—*MX3D Bridge* is a pedestrian bridge designed with generative design—complying between sustainable aspects and structural needs—and manufactured by exploiting the synergies between robotic and additive manufacturing. This is one of the first impactful examples for metal components moving from intelligent design to robotic-based production, validating the notion of the ability of such systems to move the construction sector into industrialised construction (MX3D Bridge, 2020) (Figs. 8.2 and 8.3).

Automatics monitoring for inspection—The potential of the combination between digital platform and inspection robotics is providing new opportunities for construction. This is well represented by the collaboration of Boston Dynamics and its sophisticated and movable robots SPOTWALK with HOLO BUILDER platform for the site project management controls which is revealing new digital workflows in the construction sector (HoloBuilder and Boston Dynamics Launch SpotWalk for Autonomous Reality Capture | Geo Week News | Lidar, 3D, and More Tools at the Intersection of Geospatial Technology and the Built World, 2020) (Figs. 8.4 and 8.5).

Unmanned Aerial Vehicle (UAV) for maintenance activities—UAVs could reach hazardous or high places, which is becoming a diffused

Fig. 8.2   MX3D Bridge. (Photo by Joris Laarman Lab)

Fig. 8.3   MX3D Bridge. (Photo by Adriaan de Groot)

Fig. 8.4   Spot robot for autonomous 360° photo capture. (Image courtesy of HoloBuilder)

practice with heightened expectations considering the opportunities that these technologies open to control the health of built assets. For instance, *Elios* is a UAV tool which inspects the photovoltaic (PV) panels with the aim of tracking and monitoring each cell to discover irregularities or loss of performances (Elios Aerial Thermography, 2021) (Figs. 8.6 and 8.7).

Fig. 8.5 HoloBuilder SpotWalk integration with Boston Dynamics. (Image courtesy of HoloBuilder)

Fig. 8.6 Thermography inspection of a PV plant by drone

Robotics arm in construction phase—*MULE* is a construction robot, flexible, portable, job-site ready lift assist which reduces time for lifting activities by 80% (MULE Lifting System | R.I. Lampus, 2021). ROB-Keller System AG have designed Robotic brickwork, *Rob*, to control the positioning of the masonry entirely positioned and controlled by the robotic arm. *Rob* allows to build walls even with shapes in compliance with

Fig. 8.7 Wesii digital platform: RGB view with colored anomaly classification

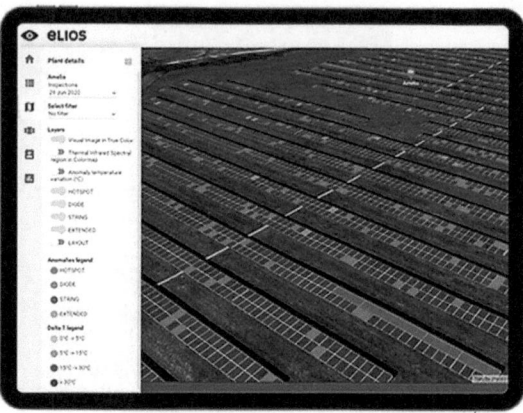

the calculations and resistance simulations made in the design phase (Robotic Brickwork, 2021).

Vehicles for construction phase—*HX2* is an autonomous and electric load carrier that can move heavy construction components. It has a vision system that allows the robot to detect humans and obstacles (Volvo CE Unveils the next Generation of Its Electric Load Carrier Concept, 2020).

Exoskeleton—*Eksovest* is an upper-body exoskeleton that supports arms during lifting activities (Exoskeletons Trialled on UK Construction Sites, 2021). *Exopush, developed by* Colas, is an exoskeleton designed to give power assistance to operatives working leveling with a rake. The exoskeleton improves the worker posture by reducing the stress movement of 30% (Colas Introduces the Exopush Exoskeleton to the UK, 2021). *G*-Ekso bionics has developed a robot which is able to hold heavy tools on aerial work platforms like scissor lifts and to standard scaffolding (EksoZeroG—Zero Gravity Tool Assistance, 2021).

Integrated solution—*Hephaestus*—A H2020 co-funded project has designed an IER tool for the installation of prefabricated building envelopes (Elia et al., 2018; Highly AutomatEd PHysical Achievements and PerformancES Using Cable RoboTs Unique Systems | HEPHAESTUS Project | Fact Sheet | H2020 | CORDIS | European Commission, 2020). The Hephaestus robot is composed of a cable-driven parallel robot (CDPR) and a modular End-Effector kit (MEE) which host tools and devices for the bracket positioning and façade modules installation. This robot expects in the next few years to provide a market autonomous solution for on-site tasks for installation of prefabricated envelopes, focusing on highly risky and critical construction tasks. The long-term purpose is to adopt Hephaestus not only for the installation stage but also for operations of maintenance and façade module replacement (Figs. 8.8 and 8.9).

Fig. 8.8 Cable-driven parallel robot installed in the demo building. (Credit Alex Iturralde)

Fig. 8.9 Hephaestus details during façade installation. (Credit Alex Iturralde)

8.5   IER FOR THE RENOVATION PHASE

In Europe more than 70% of the building stock was built before the 1970s and suffers from poor energy performance. Renovation is a key strategy to reduce the energy impact and the carbon footprint of buildings. The European Commission's target is to retrofit at least 3% of the building stock market by 2030. The retrofitting intervention involves changing in the building configuration to improve the energy performance while maintaining the occupant's comfort (Green Building 101, 2014). In this scenario construction automation and robotics can accelerate retrofitting interventions. For example, robotics applications support the existing workforce with on-site activities, which are currently based on crafts-oriented processes (Tellado, 2019). However, current key advantages of using robotics in retrofitting projects are focused on building data collection especially for the planning and design phase such as:

- Data collection regarding current building dimensions and shapes (survey). The utilisation of robotics as UAVs allows to collect accurate data in a reduced amount of time.
- Data collection regarding current building energy consumption by analyzing current building energy data, identifying areas with energy wastages, and understanding building energy use.

Robotics applications play a crucial role in addressing the challenges of building energy retrofit (Mantha et al., 2018). Accurate measurements, real time, and instant transfer of data can be integrated in the Building Information Modeling (BIM)[1] and exploited by relevant IER operations. A generic framework could be developed to support the data collected to arrive at an optimal building retrofit decision (e.g., most economical and most energy saving). Some examples are *Bertim* (*Refurbishment Solutions | STUNNING*), which is a H2020 project that aimed to enhance a building retrofitting intervention by integrating automation applications in the process, and *Vertliner* (*VERTLINER*)—an application-focused autonomous UAV that navigates inside the building, acquiring precise 3D data, images, or videos—to inform and update several layers of digital twin models and BIM representing the indoor environment.

---

[1] Chapter 3 in this book present BIM in more detail.

8.6   CHALLENGES AND BARRIERS

Despite the advantages and benefits of IER, the construction industry has faced several challenges and barriers with their adoption as summarised in Table 8.6.

Table 8.6   Challenges for adopting IER in the construction industry

Challenge/ barriers	Description
High cost of capital	While use of IER can improve productivity and reduce the labour cost, it requires high capital cost, which is not affordable for the majority of construction companies that are small and medium size (Davila Delgado et al., 2019; Llale et al., 2019).
Unstructured and dynamic nature of construction sites	Unique and unstructured nature of the construction environment, dynamics of existing objects, and ambient conditions of construction sites (e.g., adverse weather conditions and existence of dust) have been major barriers for on-site applications of IER, limiting their mobility and communication (Ardiny et al., 2015; Carra et al., 2018).
Hazards of human-robot interactions	In the current state of the construction industry, fully automated construction is a long-term goal (Czarnowski et al., 2018) and integration of human and robot is imperative (Brosque et al., 2020). Interaction of human and robots in the construction industry is a major challenge because it is fraught with safety issues such as collision and distracting workers (McCabe et al., 2017). Ensuring a safe work environment for human-robot collaboration requires development of a formal safety standard (Liang et al., 2021) and a high cost for implementing safety measures (Davila Delgado et al., 2019).
Training and skills	Lack of continuous training and the required time and cost of training construction workers to operate and collaborate with IER are the main challenges of efficiently and safely using IER in the construction industry (Davila Delgado et al., 2019; Wang et al., 2021).
Cybersecurity[a]	Cybersecurity is a major concern for IER systems. Cybersecurity threats such as malicious misuse of the robots via cyber-attacks can cause serious financial losses and safety hazards to humans (Clark et al., 2017; Yaacoub et al., 2022).

[a]Chapter 9 in this book provides a more extensive discussion on cybersecurity and privacy considerations for deep renovation

8.7   FRAMEWORKS FOR ASSESSING
AND IMPLEMENTING IER

A systematic approach to guide IER implementation is still missing in the construction sector (Hu et al., 2021; Pan et al., 2018). This section proposes a preliminary framework of indicators for assessing the advantages of using IER for buildings based on the current construction needs. The framework is designed for construction companies interested in evaluating whether robotic applications facilitate their planned tasks according to specific tasks' indicators. Using the selected metrics, the framework compares between the current manually handled tasks with the ones achievable by the adoption of a selected robotic technology. Hence, a quantitative ranking is used for the different tasks assigning a score for key macro indicators (quality, safety, time and cost) with the following scores:

- "-2" The robotic adoption hugely worsens task's indicators
- "-1" The robotic adoption worsens task's indicators
- "0" The robotic adoption does not affect task's indicators
- "+1" The robotic adoption improves task's indicators
- "+2" The robotic adoption hugely improves task's indicators

The total of all scores is a preliminary result to evaluate the IER for the selected activity: if the total score is positive, IER could facilitate the construction work, and if the total score is negative, IER will not improve the construction work.

The assessment framework is a preliminary decision support tool to facilitate the evaluation about advantages for IER adoption. More detailed investigation will need to be implemented to boost IER technologies adoption, especially once more solutions are available on the market. At this stage, the proposed framework can be considered an early-stage tool for navigating the advantages of emerging IER applications in the construction industry (Table 8.7).

Table 8.7 Framework of indicators for assessing the advantages of using IER for buildings

Frameworks for assessing and implementing construction automation and robotics

Building needs	Macro indicators and manual personnel managed						Use of robotics			
Tasks	**1. Quality** **Pitfall**	**Metric**	**2. Safety** **Pitfall**	**Metric**	**3. Time and cost** **Pitfall**	**Metric**	**1**	**2**	**3**	**Tot**
Planning and designing phase	High number of actors in construction site	No of errors	Involvement of several actors in construction safety	No of hours for safety planning	High time-consuming	No of not valuable hours of work				
Designing in retrofitting intervention	Gap between existing building and survey	Level of accuracy	Existing building survey in dangerous place	Percentage of accidental/ fatal injuries	Gap between design and as-built	No of hours for rework				
Monitoring and inspection	Reliability of monitoring data	Accuracy of the data	Inspection in dangerous place	Percentage of accidental/ fatal injuries	High number of actors in construction site	No of errors and saving in labour cost				
Repetitive and heavy tasks	Loss of attention	No of errors	High number of long-term health injuries	Percentage of long-term health injuries	High number of long-term health injuries	Percentage of long-term health injuries				
High specialised activities	Shortage of skilled workers	No of errors	High time-consuming	No of hours for training	High wage for skilled workers	Saving in labour cost				
On-site activities in hazardous location	Low level of accuracy	No of hours for rework	Dangerous place activities	Percentage of accidental/ fatal injuries	Exposed activities	No of hours of human rope				
Workers' safety training	Low level of accuracy	No of hours for rework	High time-consuming	No of hours for training	High time-consuming	No of hours for training				
Building components manufacturing	Low level of accuracy	No of hours for rework	Loss of attention due to repetitive tasks	Percentage of accidental/ fatal injuries	High amount of waste materials	Percentage of waste cost				
										TOT

8.8   CONCLUSION

There is emerging evidence that IER can benefit on-site and off-site construction operations. However, there are some challenges and barriers to overcome. From a contractor-side, economic factors including the high capital costs along with the costs pertaining to training and upskilling workers to operate IER are the main challenges. The nature of construction sites, which is generally unstructured, complex, and dynamic, entails further safety and operational challenges for using IER. Moreover, inadequate digitalisation levels within the construction industry limit the utilisation of IER. Tools for comparing traditional methods with advanced IER technologies are lacking in the construction industry. To contribute to these important gaps, this chapter classified the application of IER, reviewed key emerging applications and technologies, and proposed a framework to help assess the feasibility of implementing IER in construction. While some challenges to the adoption of IER are likely to persist in the short and mid-term, the emerging opportunities opened by IER have started to offer evidence about their disruptive nature and positive impact to quality, safety, and productivity in this key industry.

REFERENCES

Ardiny, H., Witwicki, S., & Mondada, F. (2015). Construction automation with autonomous mobile robots: A review. *2015 3rd Rsi International Conference on Robotics and Mechatronics (Icrom)*, 418–424.

Bogue, R. (2008). Cutting robots: A review of technologies and applications. *Industrial Robot-an International Journal*, 35(5), 390–396. https://doi.org/10.1108/01439910810893554

Brissi, S. G., Chong, O. W., Debs, L., & Zhang, J. S. (2022). A review on the interactions of robotic systems and lean principles in offsite construction. *Engineering Construction and Architectural Management*, 29(1), 383–406. https://doi.org/10.1108/Ecam-10-2020-0809

Brosque, C., Galbally, E., Khatib, O., & Fischer, M. (2020). Human-robot collaboration in construction: Opportunities and challenges. *2nd International Congress on Human-Computer Interaction, Optimization and Robotic Applications (Hora 2020)*, 338–345.

Carra, G., Argiolas, A., Bellissima, A., Niccolini, M., & Ragaglia, M. (2018, July 22). Robotics in the construction industry: State of the art and future opportunities. *34th International Symposium on Automation and Robotics in Construction*, Taipei, Taiwan. https://doi.org/10.22260/ISARC2018/0121

Chea, C. P., Bai, Y., Pan, X. B., Arashpour, M., & Xie, Y. P. (2020). An integrated review of automation and robotic technologies for structural prefabrication and construction. *Transportation Safety and Environment*, 2(2), 81–96. https:// doi.org/10.1093/tse/tdaa007

Choi, T. N., Chan, D. W., & Chan, A. P. (2011). Perceived benefits of applying Pay for Safety Scheme (PFSS) in construction–A factor analysis approach. *Safety Science*, 49(6), 813–823.

Chu, B., Jung, K., Lim, M. T., & Hong, D. (2013). Robot-based construction automation: An application to steel beam assembly (Part I). *Automation in Construction*, 32, 46–61. https://doi.org/10.1016/j.autcon.2012.12.016

CITB. (2021). *Construction skills network—UK 2021–2025*. https://www.citb. co.uk/media/b4fpu2hg/uk_summary.pdf

Clark, G. W., Doran, M. V., & Andel, T. R. (2017). Cybersecurity issues in robotics. *2017 Ieee Conference on Cognitive and Computational Aspects of Situation Management (Cogsima)*. WOS:000403393400021.

Colas introduces the Exopush Exoskeleton to the UK. (2021). *Colas*. https:// www.colas.co.uk/colas-news/colas-introduces-the-exopush-exoskeleton-to-the-uk

Czarnowski, J., Dąbrowski, A., & Macias, M. (2018). *Technology gaps in human-machine interfaces for autonomous construction robots*. Elsevier.

Davila Delgado, J. M., Oyedele, L., Ajayi, A., Akanbi, L., Akinade, O., Bilal, M., & Owolabi, H. (2019). Robotics and automated systems in construction: Understanding industry-specific challenges for adoption. *Journal of Building Engineering*, 26, 100868. https://doi.org/10.1016/j.jobe.2019.100868

EksoZeroG—Zero Gravity Tool Assistance. (2021). *Ekso Bionics*. https://eksobionics.com/eksozerog/

Elia, L., Alonso, R., & Cañada, J. (2018). HEPHAESTUS—Highly automated physical achievements and performances using cable robots unique systems. *Proceedings*, 2(15), 558. https://doi.org/10.3390/proceedings2150558

Elios Aerial Thermography. (2021). *Wesii*. https://www.wesii.com/elios-aerial-thermography?lang=it

European Construction Sector Observatory. (2021). *Digitalisation in the construction sector*.

Eversmann, P., Gramazio, F., & Kohler, M. (2017). Robotic prefabrication of timber structures: Towards automated large-scale spatial assembly. *Construction Robotics*, 1(1), 49–60.

Executive, U. H. and S. (2021). *Construction statistics in Great Britain, 2021*. https://www.hse.gov.uk/statistics/industry/construction.pdf

Exoskeletons trialled on UK construction sites. (2021). https://www.theb1m.com/ video/exoskeletons-trialled-on-uk-construction-site

FBR. (2022). *FBR, Innovation in the making*. https://www.fbr.com.au/

Green, B. (2021). *Industry faces a titanic struggle for skills. The RIBA Journal.* https://www.ribaj.com/intelligence/market-analysis-construction-skills-shortage-aging-workforce-reduced-immigration

Green Building 101: Why is energy efficiency important? | U.S. Green Building Council. (2014). https://www.usgbc.org/articles/green-building-101-why-energy-efficiency-important

Health and Safety Executive. (2021). *Health and safety at work.* https://www.hse.gov.uk/statistics/overall/hssh2021.pdf

Heimig, T., Kerber, E., Stumm, S., Mann, S., Reisgen, U., & Brell-Cokcan, S. (2020). Towards robotic steel construction through adaptive incremental point welding. *Construction Robotics, 4*(1), 49–60.

Highly automatEd PHysical Achievements and PerformancES using cable roboTs Unique Systems | HEPHAESTUS Project | Fact Sheet | H2020 | CORDIS | European Commission. (2020). https://cordis.europa.eu/project/id/732513/it

HoloBuilder and Boston Dynamics launch SpotWalk for autonomous reality capture | Geo Week News | Lidar, 3D, and more tools at the intersection of geospatial technology and the built world. (2020). https://www.geoweeknews.com/news/holobuilder-and-boston-dynamics-launch-spotwalk-for-autonomous-reality-capture

Hu, R., Iturralde, K., Linner, T., Zhao, C., Pan, W., Pracucci, A., & Bock, T. (2020). A simple framework for the cost–benefit analysis of single-task construction robots based on a case study of a cable-driven facade installation robot. *Buildings, 11*(1), 8. https://doi.org/10.3390/buildings11010008

Hu, R., Iturralde, K., Linner, T., Zhao, C., Pan, W., Pracucci, A., & Bock, T. (2021). A simple framework for the cost–benefit analysis of single-task construction robots based on a case study of a cable-driven facade installation robot. *Buildings, 11*(1), 8. https://doi.org/10.3390/buildings11010008

Ilyas, M., Khaw, H. Y., Selvaraj, N. M., Jin, Y. X., Zhao, X. G., & Cheah, C. C. (2021). Robot-assisted object detection for construction automation: Data and information-driven approach. *Ieee-Asme Transactions on Mechatronics, 26*(6), 2845–2856. https://doi.org/10.1109/Tmech.2021.3100306

Irani, Z., & Kamal, M. M. (2014). Intelligent systems research in the construction industry. *Expert Systems with Applications, 41*(4), 934–950. https://doi.org/10.1016/j.eswa.2013.06.061

Kim, M., Konstantzos, I., & Tzempelikos, A. (2020). Real-time daylight glare control using a low-cost, window-mounted HDRI sensor. *Building and Environment, 177*, 106912. https://doi.org/10.1016/j.buildenv.2020.106912

Kim, S., Moore, A., Srinivasan, D., Akanmu, A., Barr, A., Harris-Adamson, C., Rempel, D. M., & Nussbaum, M. A. (2019). Potential of exoskeleton technologies to enhance safety, health, and performance in construction: Industry perspectives and future research directions. *IISE Transactions on Occupational Ergonomics and Human Factors,* 7(3–4), 185–191. https://doi.org/10.108 0/24725838.2018.1561557

Lee, G., Kim, H. H., Lee, C. J., Ham, S. I., Yun, S. H., Cho, H., Kim, B. K., Kim, G. T., & Kim, K. (2009). A laser-technology-based lifting-path tracking system for a robotic tower crane. *Automation in Construction, 18*(7), 865–874. https://doi.org/10.1016/j.autcon.2009.03.011

Liang, C. J., Kang, S. C., & Lee, M. H. (2017). RAS: A robotic assembly system for steel structure erection and assembly. *International Journal of Intelligent Robotics and Applications, 1*(4), 459–476. https://doi.org/10.1007/s41315-017-0030-x

Liang, C.-J., Wang, X., Kamat, V. R., & Menassa, C. C. (2021). Human–robot collaboration in construction: Classification and research trends. *Journal of Construction Engineering and Management, 147*(10). https://doi.org/10.1061/(ASCE)CO.1943-7862.0002154

Lichiardopol, S. (2007). A survey on teleoperation. *Technische Universitat Eindhoven, DCT Report, 20,* 40–60.

Lin, J. J., Ibrahim, A., Sarwade, S., & Golparvar-Fard, M. (2021). Bridge inspection with aerial robots: automating the entire pipeline of visual data capture, 3D mapping, defect detection, analysis, and reporting. *Journal of Computing in Civil Engineering, 35*(2). https://doi.org/10.1061/(Asce)Cp.1943-5487.0000954

Llale, J., Setati, M., Mavunda, S., Ndlovu, T., Root, D., & Wembe, P. (2019). *A review of the advantages and disadvantages of the use of automation and robotics in the construction industry.* 197–204.

Mantha, B. R. K., Menassa, C. C., & Kamat, V. R. (2018). Robotic data collection and simulation for evaluation of building retrofit performance. *Automation in Construction, 92,* 88–102. https://doi.org/10.1016/j.autcon.2018.03.026

Martinez, J. G., Gheisari, M., & Alarcon, L. F. (2020). UAV integration in current construction safety planning and monitoring processes: Case study of a high-rise building construction project in Chile. *Journal of Management in Engineering,36*(3).https://doi.org/10.1061/(Asce)Me.1943-5479.0000761

McCabe, B. Y., Hamledari, H., Shahi, A., Zangeneh, P., & Azar, E. R. (2017). Roles, benefits, and challenges of using UAVs for indoor smart construction applications. *Computing in Civil Engineering 2017: Sensing, Simulation, and Visualization,* 349–357.

Melenbrink, N., Werfel, J., & Menges, A. (2020). On-site autonomous construction robots: Towards unsupervised building. *Automation in Construction, 119.* https://doi.org/10.1016/j.autcon.2020.103312

Melo, J., Sanchez, E., & Diaz, I. (2012). Adaptive admittance control to generate real-time assistive fixtures for a cobot in transpedicular fixation surgery. In *2012 4th IEEE Ras & Embs International Conference on Biomedical Robotics and Biomechatronics (Biorob)* (pp. 1170–1175). IEEE.

MULE Lifting System | R.I. Lampus. (2021). https://www.lampus.com/Product-MULE-Lifting-System

MX3D Bridge. (2020). MX3D. https://mx3d.com/projects/mx3d-bridge/

Outay, F., Mengash, H. A., & Adnan, M. (2020). *Applications of unmanned aerial vehicle (UAV) in road safety, traffic and highway infrastructure management: Recent advances and challenges.* Elsevier. https://doi.org/10.1016/j.tra.2020.09.018

Pan, M., Linner, T., Pan, W., Cheng, H., & Bock, T. (2018). A framework of indicators for assessing construction automation and robotics in the sustainability context. *Journal of Cleaner Production, 182,* 82–95. https://doi.org/10.1016/j.jclepro.2018.02.053

Pan, W., Iturralde Lerchundi, K., Hu, R., Linner, T., & Bock, T. (2020, October 14). Adopting off-site manufacturing, and automation and robotics technologies in energy-efficient building. *37th International Symposium on Automation and Robotics in Construction,* Kitakyushu, Japan. https://doi.org/10.22260/ISARC2020/0215

Reichenbach, S., & Kromoser, B. (2019). State of practice of automation in precast concrete production. *Materials Today Nano, 6.* https://doi.org/10.1016/j.jobe.2021.102527

Ribeirinho, M. J., Mischke, J., Strube, G., Sjödin, E., Blanco, J. L., Palter, R., Biörck, J., Rockhill, D., & Andersson, T. (2020). *The next normal in construction: How disruption is reshaping the world's largest ecosystem.* McKinsey & Company.

Robotic Brickwork. (2021). ROB Technologies. https://rob-technologies.com/robotic-brickwork

Saidi, K. S., Bock, T., & Georgoulas, C. (2016). *Robotics in construction* (p. 27).

Tellado, N. (2019). *Prefabrication for the building energy renovation, BERTIM methodology.* 8.

U.S. Bureau of Labor. (2021). *National Census of Fatal Occupational Injuries in 2020.* https://www.bls.gov/news.release/pdf/cfoi.pdf

Volvo CE unveils the next generation of its electric load carrier concept: Volvo construction equipment. (2020). https://www.volvoce.com/global/en/news-and-events/press-releases/2017/conexpo-vegas-2017/volvo-ce-unveils-the-next-generation-of-its-electric-load-carrier-concept/

Wang, X., Liang, C. J., Menassa, C. C., & Kamat, V. R. (2021). Interactive and immersive process-level digital twin for collaborative human-robot construction work. *Journal of Computing in Civil Engineering, 35*(6). https://doi.org/10.1061/(Asce)Cp.1943-5487.0000988

Willmann, J., Knauss, M., Bonwetsch, T., Apolinarska, A. A., Gramazio, F., & Kohler, M. (2016). Robotic timber construction—Expanding additive fabrication to new dimensions. *Automation in Construction, 61,* 16–23. https://doi.org/10.1016/j.autcon.2015.09.011

Yaacoub, J. P. A., Noura, H. N., Salman, O., & Chehab, A. (2022). Robotics cyber security: Vulnerabilities, attacks, countermeasures, and recommendations. *International Journal of Information Security, 21*(1), 115–158. https://doi.org/10.1007/s10207-021-00545-8

Yilmaz, Ş. E., & Metin, B. (2020). *On-site robotic technologies in building construction.* 10.

Zimroz, R., Hutter, M., Mistry, M., Stefaniak, P., Walas, K., & Wodecki, J. (2019). *Why should inspection robots be used in deep underground mines?* (pp. 497–507).

CHAPTER 9

Cybersecurity Considerations for Deep Renovation

Muammer Semih Sonkor and Borja García de Soto

Abstract Deep renovation efforts to improve the energy performance of buildings are of paramount importance for the overall energy reduction of nations. Like other construction projects, deep renovation ones are affected by the digital transformation of the construction industry. While this transformation involves the increasing utilisation of new technologies to optimise cost, time and quality at every stage, concerns emerge about how to maintain robust cybersecurity. This chapter summarises the cyber-security research related to each deep renovation phase and provides an overview of relevant cybersecurity frameworks, standards, guidelines and codes of practice. The chapter also discusses the need for a contingency approach in deep renovation cybersecurity due to the varying require-ments of each project and organisation.

M. S. Sonkor • B. García de Soto (✉)
S.M.A.R.T. Construction Research Group, Division of Engineering,
New York University Abu Dhabi (NYUAD), Abu Dhabi, United Arab Emirates
e-mail: semih.sonkor@nyu.edu; garcia.de.soto@nyu.edu

© The Author(s) 2023   135
T. Lynn et al. (eds.), *Disrupting Buildings*, Palgrave Studies in
Digital Business & Enabling Technologies,
https://doi.org/10.1007/978-3-031-32309-6_9

Keywords BIM • Construction 4.0 • Cybersecurity • Cyber-physical systems • Digitalisation • Information technology (IT) • Internet of Things (IoT) • Operational technology (OT)

9.1   INTRODUCTION

Sustainable development constitutes one of the highest priority topics on most national agendas, and energy efficiency has a critical role in achieving sustainability targets. Buildings consume a significant amount of energy (Lynn et al., 2021); therefore, reducing the energy consumption of existing buildings can help countries achieve these targets and enhance global energy efficiency. Shnapp et al. (2013, p. 19) define deep renovation as "a renovation that captures the full economic energy efficiency potential of improvement works, with a main focus on the building shell, of existing buildings that leads to a very high-energy performance"[1]. While widely referenced, it is important to note that there is no consensus on the definition of deep renovation and the associated minimum energy reduction required.

Deep renovation can be considered a specialised subcategory of construction. It thus passes through similar stages (e.g., design, construction/retrofitting, operation and maintenance (O&M) and end of life) as with other construction projects, even though its scope involves retrofitting existing buildings rather than building one from the ground up. Therefore, technological advances in the construction industry and the concerns related to these advances are also applicable to deep renovation projects. The digitalisation that the construction industry is going through, often referred to as Construction 4.0 (Klinc & Turk, 2019), affects the information generated and used and the physical tasks performed during the construction and O&M phases (Garcíade Soto et al., 2020). While this transformation improves the cost and time efficiency of processes and construction quality, it also leads to substantial cybersecurity concerns, as with other digitalised industries. The convergence of information technology (IT) and operational technology (OT) (Harp & Gregory-Brown, 2015) further exacerbates the difficulty and complexity of addressing such concerns. Furthermore, safety issues arise due to the increasing use of OT

---

[1] Chapter 1 in this book provides a more detailed discussion on the definition of deep renovation.

to perform (e.g., autonomous excavators to handle earthworks) and monitor (e.g., autonomous site monitoring devices) site activities (Sonkor & García de Soto, 2021). As a result, the significance of providing robust cybersecurity increases during all phases of construction projects to prevent the exposure of sensitive information and any potential physical damage.

The rest of this chapter is organised as follows. Next, we discuss major types of cybercrimes that affect the construction industry and the related laws and regulations. We then outline some prominent cybersecurity standards, codes of practice and frameworks applicable to the construction industry. Following a review of relevant cybersecurity research organised by the deep renovation phase, we explain the need for a contingency approach to cybersecurity in the construction industry that takes into account the differences in projects, organisations and contexts while highlighting that there cannot be a one-size-fits-all solution for all different sizes of companies and deep renovation projects.

9.2   Cybercrimes and Cybersecurity in Construction

Increased connectivity, remote working and the increasing sophistication of malicious actors are contributing to a rise in cybercrime (FireEye, 2021). The construction sector is not insulated from this trend. As more and more buildings become reliant on remotely operated software systems and the Internet of Things, the attack surface and associated vulnerabilities and risks increase. Construction companies and their employees, specific projects and building systems have been targeted by a wide range of cyberattacks, including phishing, ransomware, denial of service, identity theft and other types of unauthorised access (Nordlocker, 2021; Korman, 2020; Turton & Mehrotra, 2021; Rashid et al., 2019). While financial gain is a common motivation for such attacks, it is not always the case. For example, in 2016, hackers launched a distributed denial of service (DDoS) attack on two residential buildings in Finland by temporarily disabling the computer systems that controlled the heating and hot water distribution systems, resulting in obvious inconvenience and distress for residents (Ashok, 2016). Unsurprisingly, governments worldwide have responded to the threat of cyberattacks. These actions include enacting new laws focusing on cybercrimes and introducing acts and regulations that define criminal offences and the related sanctions. Notwithstanding this, few are

Table 9.1 Common types of cybercrimes, examples from the construction industry and related laws and regulations

Cybercrime	Description	Examples	Related laws and regulations
Phishing	Attempting to convince the victim to reveal sensitive information through e-mail or other means of communication using technical subterfuge and social engineering techniques (Dou et al., 2017)	Turner Construction (Jones, 2016); Central Concrete Supply Co. Inc. (Jones, 2016)	• EU Directive 2019/713 on combating fraud and counterfeiting of non-cash means of payment • 18 U.S. Code § 1343—Fraud by wire, radio or television • UK Fraud Act, 2006
Infection of IT systems with malware (including ransomware and viruses)	Using malicious software such as trojans, spyware, ransomware and botnet malware to perform malicious activities. These activities and their outcomes depend on the type of malware. For example, ransomware can encrypt data, and trojans can steal confidential information (Rashid et al., 2019)	Bouygues Construction (Korman, 2020); Bam Construct (Muncaster, 2020); Grey Energy (Cherepanov, 2018); E.R. Snell Contractor, Inc. (Equipment World, 2022)	• EU Directive 2013/40 on attacks against information systems • US Computer Fraud and Abuse Act (CFAA), 1986 • UK Computer Misuse Act, 1990
Hacking (unauthorised access)	The act of accessing a cyber system without having the right and authorisation (Rashid et al., 2019)	Colonial Pipeline (Turton & Mehrotra, 2021); Oregon Construction Contractors Board (CCB) security breach (Oregon CCB, 2019)	• EU Directive 2013/40 on attacks against information systems • CFAA, 1986 • 18 U.S. Code § 1030—Fraud and related activity in connection with computers • UK Computer Misuse Act, 1990
Denial-of-service attacks	Depleting a system's computing resources (e.g., CPU, memory) to cause unavailability and inaccessibility (Rashid et al., 2019)	Valtia attack (Ashok, 2016); Ukraine power grid attack (Slowik, 2019)	

(*continued*)

Table 9.1 (continued)

Cybercrime	Description	Examples	Related laws and regulations
Identity theft or identity fraud	Stealing an individual's or business's identity information to use it for deception or fraud, usually for financial gain (US DOJ, 2020)	CCB License Fraud (Oregon.gov, 2014); Contractor License Fraud in San Diego (FOX 5 San Diego, 2015); Konecranes (Reuters, 2015)	• EU Directive 2019/713 on combating fraud and counterfeiting of non-cash means of payment • US Identity Theft and Assumption Deterrence Act, 1998 • US Identity Theft Penalty Enhancement Act, 2004 • UK Fraud Act, 2006

specifically focused on the construction industry and buildings per se. Table 9.1 summarises common cybercrimes, examples from the construction industry and related laws and regulations.

9.3   International Standards, Best Practices and Cybersecurity Frameworks

In recent years, national and international institutions have been active in producing standards and guidelines to support companies in assessing their current cybersecurity levels and setting targets for the future. While the overwhelming majority are aimed at the IT sector or firms in general, there are several codes of practice and guidelines aimed at the architecture, engineering, construction and operations (AECO) sector specifically. As modern buildings make widespread use of automation and control systems, for example, for heating, and such systems have been the target of

Table 9.2 Summary of the commonly used cybersecurity standards and procedures

Title	Published by	Main purpose and applications	Target industry	Year	Source
ISO 19650-5:2020—Information management using building information modelling—Part 5: Security-minded approach to information management	ISO	An international standard that introduces a security-minded approach for construction projects and built environments that utilises building information modelling (BIM) in their processes. It targets lowering the risk of loss or unauthorised alteration of sensitive information in built environments	AECO	2020	ISO (2020)
ISO/IEC 27001:2013—Information technology—Security techniques—Information security management systems—Requirements	ISO/IEC	A set of standards that guides companies in implementing and managing information security management systems (ISMS). It can be used either internally or by third parties to evaluate the company's capability to meet the information security requirements. It uses the "Plan-Do-Check-Act" model for structuring ISMS processes	All industries	2013	ISO/IEC (2013)
Service Organization Control (SOC) 2	AICPA	An auditing procedure that defines requirements for organisations to manage their customers' data securely. It is based on five trust principles: confidentiality, processing integrity, availability, security and privacy	All industries	2018	AICPA (2018)

(*continued*)

Table 9.2 (continued)

Title	Published by	Main purpose and applications	Target industry	Year	Source
ISA/IEC 62443—Security for Industrial Automation and Control Systems (IACSs)	ISA/ IEC	A series of standards that provides guidelines to identify and mitigate cybersecurity vulnerabilities in IACSs. It applies to all industries and critical infrastructures (CIs)	Organisations that implement IACSs	2020[a]	ISA (2020)

[a]Indicates the last publication date of an ISA/IEC 62443 Series (in this case, Part 3-2: Security risk assessment for system design)

cyberattacks, standards and guidelines for the security of such systems are also relevant. While some are industry-specific, others were designed in a generic way to cover a wide range of sectors. Some of the commonly used standards and procedures for cybersecurity are presented in Table 9.2.

In addition to standards and protocols for security and control systems, there are several codes of practice and guidelines. Some are general (for any industry), but others specifically address the construction sector. While codes of practice do not purport to replace standards, they provide guidance and support for achieving standards. Table 9.3 summarises some of the prominent codes of practice, guidelines and frameworks for cybersecurity.

9.4   Related Cybersecurity Research by Renovation Phase

To date, scholarly research has focused primarily on the advantages and potential benefits of increased digitalisation of the construction sector. In comparison, cybersecurity aspects have received less attention. There are notable exceptions. For example, Turk et al. (2022) proposed a systematic framework to address the cybersecurity problems specific to construction projects. Their framework identified cybersecurity as "the absence of the three wrongs across the four kinds of elements" (Turk et al., 2022, p. 1). The three wrongs refer to stealing, harming and lying. The four elements

Table 9.3 Summary of the commonly cite codes of practice, guidelines and frameworks for cybersecurity

Title	Published by	Main purpose and applications	Target industry	Year	Source
Cyber Security for Construction Businesses	The National Cyber Security Centre (NCSC)/ Chartered Institute of Building (CIOB)/UK Government	Provides cybersecurity guidance to small and medium-sized businesses in the construction industry and provides practical advice for each stage of construction. It helps business owners and managers understand why cybersecurity matters and advises staff responsible for IT equipment and services within construction companies on actions to take	AECO	2022	NCSC (2022)
Code of Practice: Cyber Security in the Built Environment— 2nd Edition	The Institution of Engineering and Technology (IET)	Provides guidance to the stakeholders of built environments at every stage of the buildings, from planning to disposal. It identifies the different cybersecurity threats and requirements related to each phase	AECO	2021	IET (2021)

(*continued*)

Table 9.3 (continued)

Title	Published by	Main purpose and applications	Target industry	Year	Source
Secure by Design: Improving the Cyber Security of Consumer Internet of Things Report	Department for Digital, Culture, Media & Sport/UK Government	A collection of guidelines on consumer Internet of Things (IoT) security that includes a code of practice and several other sections that discuss risks, opportunities and government actions. The code of practice, which is the central part of this report, involves thirteen suggestions prepared by experts to improve IoT cybersecurity	All sectors in the UK using IoT	2021	UK Government (2021)
Cyber Assessment Framework v3.0	NCSC	The framework has four objectives: protection against cyberattacks, management of cybersecurity risks, detection of cybersecurity incidents and minimising the impact of such incidents	All industries	2019	NCSC (2019)

(*continued*)

Table 9.3 (continued)

Title	Published by	Main purpose and applications	Target industry	Year	Source
Framework for Improving Critical Infrastructure Cybersecurity v1.1	NIST	Provides an extensive guideline for companies to develop their assessment structures using different reference documents (e.g., frameworks and standards)	CI operators; all organisations	2018	NIST (2018)
Network and Information Systems (NIS) Directive	EU	A legislative document that requires EU member states to have national cybersecurity strategies and encourages cooperation among the members to enhance the overall cybersecurity level within the Union	All sectors in the EU	2016	EU (2016)

that might be affected by such wrongful activities are material, information, person and system. After defining cybersecurity, they customised the framework to reflect construction-specific characteristics. These characteristics include the multi-stakeholder settings of projects, overlapping boundaries of different entities involved in different projects and having distinct stages (e.g., design, construction and O&M) with particular challenges.

Several studies in recent years have discussed various aspects of construction cybersecurity and suggested solutions across the construction and deep renovation life cycle. Zheng et al. (2019) stressed the lack of studies concerning the information security aspects of BIM during the design and planning phase. In order to improve confidentiality and reduce the risk of data breaches, a context-aware access control model named

CaACBIM was proposed. Mantha et al. (2021) pointed out that the sensor data collected during the commissioning phase can be altered by malicious actors (e.g., an owner with a malicious intention or a competitor). In order to address this threat, they proposed utilising an autonomous robotic system for randomised check-pointing and illustrated its feasibility with an example.

Modern construction and retrofitting make increasing use of (semi) autonomous and remote-controlled equipment (Sonkor & García de Soto, 2021). This includes the use of complex cyber-physical systems, such as industrial machinery and vehicles (e.g., cranes), exoskeletons, unmanned aerial vehicles (UAV),[2] on-site and off-site automated fabrication and additive manufacturing,[3] to name a few. Notwithstanding the pervasiveness of such equipment, a recent survey of cybersecurity research on such construction equipment by Sonkor and García de Soto (2021) revealed a paucity of studies.

Many of the construction cyberattacks identified in Table 9.1 occur in the O&M phase of construction projects, particularly in smart buildings. Pärn and Edwards (2019) presented the potential cybersecurity issues for CIs during the O&M phase and suggested using blockchain technology for data exchange and storage as a mitigation action. Several studies focused on the cybersecurity aspects of smart buildings. Wendzel et al. (2014) discussed botnets' abilities to control and monitor building automation systems (BASs) and their potential damage to the built environment. On a related topic, Mundt and Wickboldt (2016) undertook a study to identify the cyber risks, possible attackers and attack vectors related to BASs. They presented the security gaps found in two case studies to prove that additional attention is required to ensure robust BAS security. Mirsky et al. (2017) showed how air-gapped building management networks could be attacked using a compromised heating, ventilation and air conditioning (HVAC) system. Lastly, Wendzel et al. (2017) investigated the potential attacks against smart buildings and proposed solutions to protect them.

Interestingly, few studies explore the end-of-life phase of buildings and construction projects from a cybersecurity point of view. As building systems may retain sensitive data that can be exposed due to vulnerabilities, care needs to be taken to ensure suitable cybersecurity safeguards are in place.

---

[2] See Chap. 8 in this book for a more detailed discussion.
[3] See Chap. 7 in this book for a more detailed discussion.

While it is useful from a research perspective to use a phased approach to identify gaps in the literature, many actors and systems in the construction and renovation process are present across the entire life cycle, particularly as a consequence of digitisation. As such, full life-cycle approaches to cybersecurity assessment and associated research are needed. For example, Mantha and García de Soto (2019) investigated the vulnerability of different project participants and construction entities during the different phases of the life cycle of construction projects as a consequence of Construction 4.0. Their study considered potential risks and provided a basis for assessing the impact of interactions in a digital environment among different project participants. Considering the increasing use of IoT, edge computing and artificial intelligence (AI) and the likelihood that every stage of construction and deep renovation projects is expected to rely on these technologies in the near future, their cybersecurity vulnerabilities and risks require more attention (Ansari et al., 2020).

9.5   The Need for a Contingency Approach

The primary purpose of all the previously mentioned cybersecurity standards, frameworks, guidelines and academic studies is to improve the cybersecurity level of projects and organisations. However, considering the variety in functions, roles and scale differences in construction and deep renovation firms and projects, a one-size-fits-all cybersecurity approach may not be desirable or feasible. For example, public companies will have to meet specific accounting standards to ensure adequate controls are in place, and multinational firms may have to deal with a wide range of cybersecurity and data protection requirements. Similarly, specialist craft renovations are likely to have different cybersecurity requirements and demands than more generic and large-scale construction/renovation projects. Each stakeholder constitutes a different cyber risk, and each one has various cybersecurity concerns. Therefore, care needs to be taken to ensure that an appropriate cybersecurity assessment and associated controls are put in place that can accommodate the range of projects and firms that characterise the sector.

9.6   Conclusion

The integration of construction and digital technologies such as IoT, machine learning and cloud computing disrupts how construction projects are planned, constructed and operated, making the construction industry and buildings easy targets. At the same time, the sophistication and volume of cyberattacks are increasing. As an inevitable consequence, maintaining robust cybersecurity becomes an everyday challenge. Deep renovation projects face the same hurdles as any other construction project when it comes to protecting sensitive information and maintaining safety. This chapter provides an overview of the cybersecurity efforts in the construction industry and deep renovation and presents relevant frameworks, standards, codes of practice and research. Furthermore, it discusses the need for a contingency approach while considering the cybersecurity requirements of deep renovation projects and the firms that deliver them. There is no silver bullet in cybersecurity. Cybersecurity considerations and related actions should be an indispensable part of deep renovation projects from planning to the end of life, taking into account the needs of all stakeholders.

Acknowledgements The authors would like to thank the Center for Cyber Security at New York University Abu Dhabi (CCS-NYUAD) for the support provided for this study.

REFERENCES

AICPA. (2018). *SOC 2*. https://www.aicpa.org/cpe-learning/publication/soc-2-reporting-on-an-examination-of-controls-at-a-service-organization-relevant-to-security-availability-processing-integrity-confidentiality-or-privacy

Ansari, M. S., Alsamhi, S. H., Qiao, Y., Ye, Y., & Lee, B. (2020). Security of distributed intelligence in edge computing: Threats and countermeasures. In T. Lynn, J. G. Mooney, B. Lee, & P. T. Endo (Eds.), *The cloud-to-thing continuum* (pp. 95–122). Springer International Publishing. https://doi.org/10.1007/978-3-030-41110-7_6

Ashok, I. (2016). Hackers leave Finnish residents cold after DDoS attack knocks out heating systems. *Yahoo News*. https://sg.news.yahoo.com/hackers-leave-finnish-residents-cold-105147593.html

Cherepanov, A. (2018). *Greyenergy—A successor to Blackenergy*. ESET. https://www.welivesecurity.com/wp-content/uploads/2018/10/ESET_Grey Energy.pdf

García de Soto, B., Georgescu, A., Mantha, B. R. K., Turk, Ž., & Maciel, A. (2020). Construction cybersecurity and critical infrastructure protection: Significance, overlaps, and proposed action plan. *Preprints 2020*. https://doi.org/10.20944/preprints202005.0213.v1

Dou, Z., Khalil, I., Khreishah, A., Al-Fuqaha, A., & Guizani, M. (2017). Systematisation of Knowledge (SoK): A systematic review of software-based web phishing detection. *IEEE Communication Surveys and Tutorials, 19*(4). https://doi.org/10.1109/COMST.2017.2752087

Equipment World. (2022). Hacked: Construction contractor E.R. Snell shares how to bounce back from a cyberattack. *Equipment World.* https://www. equipmentworld.com/business/article/15290439/how-to-protect-your-construction-business-from-cyberattacks

EU. (2016). *Directive (EU) 2016/1148 of the European Parliament and of the Council of 6 July 2016 concerning measures for a high common level of security of network and information systems across the Union.* Official Journal of the European Union. http://data.europa.eu/eli/dir/2016/1148/oj

FireEye. (2021). *M-Trends 2021.* https://content.fireeye.com/m-trends/ rpt-m-trends-2021

FOX 5 San Diego. (2015). Construction contractor accused of fraud, identity theft. *FOX 5 San Diego.* https://fox5sandiego.com/news/construction-contractor-accused-in-identify-theft-scam/

Harp, D. R., & Gregory-Brown, B. (2015). IT / OT convergence bridging the divide. *NexDefense.* https://ics.sans.org/media/IT-OT-Convergence-NexDefense-Whitepaper.pdf

IET. (2021). *Code of Practice: Cyber security in the built environment—2nd edition.* https://electrical.theiet.org/guidance-codes-of-practice/publications-by-category/cyber-security/code-of-practice-cyber-security-in-the-built-environment-revised-second-edition/

ISA. (2020). *Quick Start Guide: An overview of ISA/IEC 62443 Standards.* Security of Industrial Automation and Control Systems, International Society of Automation (ISA), Global Cybersecurity Alliance. https://gca.isa.org/ hubfs/ISAGCA Quick Start Guide FINAL.pdf

ISO. (2020). *ISO 19650-5:2020 Organisation and digitisation of information about buildings and civil engineering works, including building information modelling (BIM)—Information management using building information modelling—Part 5.* https://www.iso.org/standard/74206.html

ISO/IEC. (2013). *ISO/IEC 27001:2013—Information technology—Security techniques -Information security management systems—Requirements.* https:// www.iso.org/standard/54534.html

Jones, K. (2016). *Data breaches, cybersecurity, and the construction industry.* Construct Connect (Blog). https://www.constructconnect.com/blog/data-breaches-cyber-security-construction-industry

Klinc, R., & Turk, Ž. (2019). Construction 4.0—Digital transformation of one of the oldest industries. *Economic and Business Review, 21*(3), 393–410. https:// doi.org/10.15458/ebr.92

Korman, R. (2020). *Bouygues construction unit gradually recovering after ransomware attack*. Engineering News-Record (ENR). https://www.enr.com/articles/48637-bouygues-construction-unit-gradually-recovering-after-ransomware-attack

Lynn, T., Rosati, P., Egli, A., Krinidis, S., Angelakoglou, K., Sougkakis, V., Tzovaras, D., Kassem, M., Greenwood, D., & Doukari, O. (2021). RINNO: Towards an open renovation platform for integrated design and delivery of deep renovation projects. *Sustainability, 13*(11). https://doi.org/10.3390/su13116018

Mantha, B. R. K., & García de Soto, B. (2019). Cyber security challenges and vulnerability assessment in the construction industry. *Creative Construction Conference*, 29–37. https://doi.org/10.3311/ccc2019-005

Mantha, B. R. K., García de Soto, B., & Karri, R. (2021). Cyber security threat modeling in the AEC industry: An example for the commissioning of the built environment. *Sustainable Cities and Society, 66*, 102682. https://doi.org/10.1016/j.scs.2020.102682

Mirsky, Y., Guri, M., & Elovici, Y. (2017). HVACKer: Bridging the air-gap by attacking the air conditioning system. *ArXiv.org*. https://doi.org/10.48550/ARXIV.1703.10454

Muncaster, P. (2020). COVID19 hospital construction firms hit by cyber-Attacks. *Infosecurity Magazine*. https://www.infosecurity-magazine.com/news/covid19-hospital-construction/

Mundt, T., & Wickboldt, P. (2016). Security in building automation systems—A first analysis. *2016 International Conference on Cyber Security and Protection of Digital Services, Cyber Security 2016*. https://doi.org/10.1109/CyberSecPODS.2016.7502336

NCSC. (2019). *Cyber Assessment Framework v3.0*. https://www.ncsc.gov.uk/files/NCSC_CAF_v3.0.pdf

NCSC. (2022). *Cyber security for construction businesses*. NCSC. https://www.ncsc.gov.uk/guidance/cyber-security-for-construction-businesses

NIST. (2018). *Framework for improving critical infrastructure cybersecurity v1.1*. https://doi.org/10.6028/NIST.CSWP.04162018

Nordlocker. (2021). *Top industries hit by ransomware*. Nordlocker. https://nordlocker.com/recent-ransomware-attacks/

Oregon CCB. (2019). *Construction contractors board takes steps to stop data and security breach, inform contractors*. https://www.oregon.gov/CCB/Documents/pdf/JUSTICE-9596167-v1-CCB_-_Data_Disclosure_News_Release.pdf

Oregon.gov. (2014). *Con artist goes to prison for using stolen CCB license number*. Oregon.Gov. https://www.oregon.gov/CCB/news/Pages/stolen CCBlicensenumber.aspx

Pärn, E., & Edwards, D. (2019). Cyber threats confronting the digital built environment: Common data environment vulnerabilities and block chain deterrence. *Engineering Construction and Architectural Management*, *26*(2), 245–266. https://doi.org/10.1108/ECAM-03-2018-0101

Rashid, A., Chivers, H., Danezis, G., Lupu, E., & Martin, A. (2019). *The Cyber Security Body of Knowledge (CyBOK) v1.0*. University of Bristol. https://www.cybok.org/

Reuters. (2015). *Finland's Konecranes says subsidiary hit by fraud*. Reuters. https://www.reuters.com/article/konecranes-fraud-idUSFWN10P0 5K20150814

Shnapp, S., Sitjà, R., & Laustsen, J. (2013). *What is a deep renovation definition?* https://www.gbpn.org/wp-content/uploads/2021/06/08.DR_TechRep. low_pdf

Slowik, J. (2019). *Crashoverride: Reassessing the 2016 Ukraine electric power event as a protection-focused attack*. https://www.dragos.com/wp-content/uploads/CRASHOVERRIDE.pdf

Sonkor, M. S., & García de Soto, B. (2021). Operational technology on construction sites: A review from the cybersecurity perspective. *Journal of Construction Engineering and Management*, *147*(12). https://doi.org/10.1061/(ASCE) CO.1943-7862.0002193

Turk, Ž., García de Soto, B., Mantha, B. R. K., Maciel, A., & Georgescu, A. (2022). A systemic framework for addressing cybersecurity in construction. *Automation in Construction*, *133*(January), 103988. https://doi.org/10.1016/j. autcon.2021.103988

Turton, W., & Mehrotra, K. (2021). Hackers breached Colonial Pipeline using compromised password. *Bloomberg*. https://www.bloomberg.com/news/articles/2021-06-04/hackers-breached-colonial-pipeline-using-compromised-password

UK Government. (2021). *Secure by design*. UK Government. https://www.gov.uk/government/collections/secure-by-design

US DOJ. (2020). *Identity theft*. Justice.Gov. https://www.justice.gov/criminal-fraud/identity-theft/identity-theft-and-identity-fraud

Wendzel, S., Tonejc, J., Kaur, J., & Kobekova, A. (2017). Cyber security of smart buildings. In H. Song, G. A. Fink, & S. Jeschke (Eds.), *Security and privacy in cyber-physical systems: Foundations, principles, and applications*. John Wiley & Sons Ltd.

Wendzel, S., Zwanger, V., Meier, M., & Szlósarczyk, S. (2014). Envisioning smart building botnets. *Lecture Notes in Informatics (LNI), Proceedings—Series of the Gesellschaft Fur Informatik (GI)*, *P-228* (pp. 319–329).

Zheng, R., Jiang, J., Hao, X., Ren, W., Xiong, F., & Zhu, T. (2019). CaACBIM: A context-aware access control model for BIM. *Information*, *10*(2), 47. https://doi.org/10.3390/info10020047

Financing Building Renovation: Financial Technology as an Alternative Channel to Mobilise Private Financing

\

Mark Cummins, Theo Lynn, and Pierangelo Rosati

Abstract Access to capital is one of the key barriers for deep renovation. This chapter presents the potential advantages and benefits that financial technology (FinTech) solutions such as crowdfunding and blockchain-based solutions such as tokenisation and smart contracts can provide to building owners and construction companies in terms of financing. Future avenues for research in this space are also presented.

M. Cummins
Strathclyde Business School, University of Strathclyde, Glasgow, UK
e-mail: mark.cummins@strath.ac.uk

T. Lynn
Irish Institute of Digital Business, DCU Business School,
Dublin City University, Dublin, Ireland
e-mail: theo.lynn@dcu.ie

P. Rosati (✉)
J.E. Cairnes School of Business and Economics, University of Galway,
Galway, Ireland
e-mail: pierangelo.rosati@universityofgalway.ie

© The Author(s) 2023   153
T. Lynn et al. (eds.), *Disrupting Buildings*, Palgrave Studies in
Digital Business & Enabling Technologies,
https://doi.org/10.1007/978-3-031-32309-6_10

Keywords FinTech • Crowdfunding • Blockchain • Smart contracts • Alternative finance

10.1   Introduction

Moves towards a long-term net zero emissions objective are complex and multifaceted. One part of this global picture that needs to be addressed effectively is the high level of energy inefficiency amongst a high proportion of buildings globally. For the EU, it was estimated, for instance, that (as of 2011) approximately 75% of the building stock in the EU required some form of energy efficiency upgrade in the form of retrofitting and renovation (Economidou et al., 2011). The Energy Efficiency Directive (Directive 2012/27/EU) of 2012 has been a key policy response by the EU to set the foundations for a significant programme of building renovation.[1] This legislation was partially revised in 2018. However, the European Commission has now commenced a process of overhauling the entire Energy Efficiency Directive,[2] seeking to leverage the Renovation Wave strategy announced in 2020.[3] This latter strategy aims to double annual energy renovation rates in the next 10 years. As well as reducing emissions, these renovations will enhance quality of life for people living in and using the buildings, and should create many additional green jobs in the construction sector.

Feedback in the Open Consultation on the Renovation Wave suggested that lack of or limited resources to finance building renovation is one of the most important barriers to building renovations. These barriers include a lack of financial incentives, access to mainstream financing products, and funding for publicly owned buildings. In response, ensuring adequate and well-targeted funding is central to the EU Renovation Wave strategy. Despite this, while the European Commission highlights the need for greater adoption of digital and innovative technologies in the construction sector and identifies specific digital tools and technologies, it is silent on financial technologies and how they might reduce barriers to building and renovation finance.

---

[1] https://eur-lex.europa.eu/LexUriServ/LexUriServ.do?uri=OJ:L:2012:315:000 1:0056:en:PDF.

[2] https://ec.europa.eu/info/news/commission-proposes-new-energy-efficiency-directive-2021-jul-14_en.

[3] https://energy.ec.europa.eu/topics/energy-efficiency/energy-efficient-buildings/renovation-wave_en.

Against this backdrop, this chapter will explore the financing of building renovation and how innovations in the financial technology (FinTech) space may serve to mobilise more private financing. The remainder of this chapter is organised as follows. Following a summary of recent literature on financing building renovation, we define and outline key FinTech concepts and technologies. We then explore two of the most prominent FinTech solutions—crowdfunding and blockchain-based solutions.

10.2 DEEP RENOVATION FINANCING: KEY TERMS AND CONCEPTS

Economidou et al. (2019) provide an overview of the main financing instruments available to support energy renovations in the EU (Fig. 10.1). These are categorised by type of financing instrument, spanning (1) non-repayable rewards, (2) debt financing, and (3) equity financing, and by market saturation, spanning (1) traditional and well-established, (2) tested and growing, and (3) new and innovative financing mechanisms. A brief definition for each instrument is provided in Table 10.1 with other key terms and concepts used in this chapter.

Kunkel (2015) sets out the barriers to traditional investment in building renovation as follows: (1) upfront investment and the bankability of projects; (2) information asymmetry; (3) the quality of the on-site implementation and the trust in local partners and companies; and (4) split incentives and uncompensated benefits. These barriers have become even more significant following the COVID-19 pandemic due to an

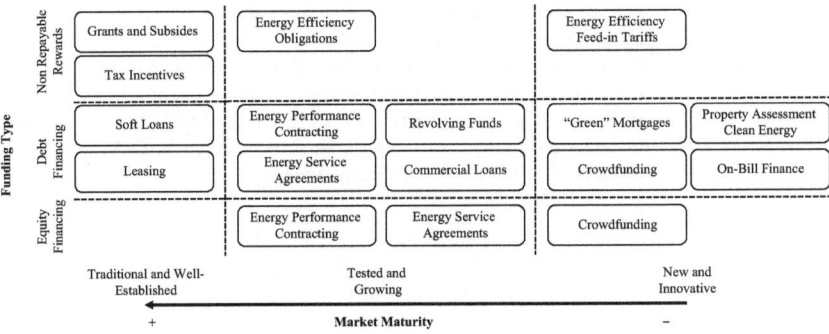

Fig. 10.1 Financing landscape in the EU for energy renovations according to market maturity and type. (Adapted from Economidou et al., 2019)

Table 10.1   Deep renovation financing key terms and concepts

Financing instrument	Definition
Blockchain	A decentralised, transactional database that enables validated, tamper-resistant transactions across a large number of participants (i.e., nodes) in a network (Glaser, 2017; Beck et al., 2017). It is the technology underpinning Bitcoin (Nakamoto, 2008), but its applications extend beyond digital currencies (Rosati & Čuk, 2019).
Commercial loans	Loans provided by commercial banks that are issued through standardised project appraisal and loan processing processes (Economidou et al., 2019). As such, they reduce uncertainties regarding access to capital and reduce transaction costs.
Crowdfunding	An open call, typically through the Internet for the provision of financial resources from a group of individuals or organisations (Belleflamme et al., 2014). In the context of building renovation, these calls typically aim to attract funding from a large number of either retail or institutional investors in exchange for a share of the property or for future revenue streams in the form of interest and principal repayments.
Energy efficiency feed-in tariffs	An instrument that aims to reduce energy use through a reward-based system (Economidou et al., 2019). While relatively simple to implement, it is typically based on a fixed price system which may ultimately favour cheap energy efficiency interventions (Eyre, 2013).
Energy efficiency obligations	Market-based instruments that can be put in place by governments to achieve energy savings through investments obligations placed on energy companies (Economidou et al., 2019).
Energy performance contracting (EPC)	A contract between an energy services company (ESCO) and a client whereby the ESCO is responsible for completing a renovation project and to deliver energy efficiency improvements on a given building owned by the client and it uses the costs savings generated by energy efficiency measures implemented to repay the costs of the project over a given time period (Lee et al., 2015).
Energy service agreements (ESA)	Represent a variant of EPC "that involve integrated financing measures, backed by a long-term performance guarantee" (Brown et al., 2022, p. 7). In this type of contracts, the ESCO bears both the financial and performance risk of the project. As such, ESAs are particularly attractive for building owners.

(*continued*)

Table 10.1 (continued)

Financing instrument	Definition
Grants and subsides	Grants represent a direct monetary contribution towards a building renovation project and serve as direct investment subsidies which may partially or fully cover the cost of the renovation. Grants and subsides are typically provided by government agencies and, as such, rely on limited resources and cannot represent a sustainable solution or support massive market uptake programmes (Economidou et al., 2019).
Green mortgages	Loans provided by commercial banks and other credit institutions that provide borrowers with the opportunity to finance the cost of energy-efficient upgrades and to benefit from preferential mortgage terms (e.g., better borrowing terms, higher debt-to-income ratios) (Economidou et al., 2019).
Leasing	A lease can be defined as a contract between the owner of an asset (lessor) and the user of such an asset (lessee), whereby the lessor provides the asset for use by the lessee for a certain time period in exchange for a payment with an understanding that, at the end of such period, the asset will either be returned to the lessor, purchased in full by the lessee for a pre-defined amount, or disposed as outlined in the contract (Oracle, 2016). In the context of building renovation, leasing contracts may incorporate clauses whereby the lessor and the lessee take on specific obligations with regard to the sustainable operation and occupation of a given property (Kaplow, 2008). These include, for example, the implementation of energy efficiency measures and waste reduction.
On-bill finance	A financing mechanism that reduces the upfront cost of energy renovation projects by linking repayments to the utility bill. As such, it allows customers to pay back the cost of the investment over time. These mechanisms can be promoted by local governments as well as utility companies (Economidou et al., 2019).
Property assessment clean energy	This instrument aims to finance energy renovations through the use of specific bonds issued by municipal governments to investors. The funds raised through the sale of these bonds are used to provide loans to building owners who want to implement energy renovations in either residential or commercial buildings. The loans are typically repaid over 15–20 years via an annual assessment on property tax bills (Economidou et al., 2019).

(*continued*)

Table 10.1 (continued)

Financing instrument	Definition
Revolving funds	An energy efficiency revolving fund provides financing and related services to its clients to facilitate energy efficiency investments (Lukas, 2018). These funds are designed to be self-sustainable as a portion of the savings generated by supported investments is used to replenish in part the fund therefore allowing for reinvestment in future projects (Lukas, 2018).
Security Token Offering (STO)	Security Token Offerings are regulated token offerings whereby the token issuer raises capital by selling to qualified investors crypto tokens that are defined as securities (Lynn & Rosati, 2021).
Smart contracts	"Self-executing electronic instructions drafted in computer code" (O'Shields, 2017, p. 179). More specifically, blockchain-based smart contracts are signed by the parties involved using cryptographic security, are stored on the blockchain, and self-execute the stipulations of an agreement when predetermined conditions are met (O'Shields, 2017).
Soft loans	Government-supported loans which may be offered at below market interest rates or allow for longer repayment periods (Karakosta et al., 2021).
Tax incentives	Aim to promote building renovation by reducing the cost of the energy efficiency improvement through reduced taxes for households and organisations (Economidou et al., 2019). Tax incentives can be designed in a number of ways such as accelerated depreciation, tax exemptions, income tax or VAT reduction, and so on.
Tokenisation	Tokenisation is one of the main applications enabled by blockchain which allows users to digitise tangible and intangible assets. Each token represents a certain share of an asset's ownership and it can be recorded and exchanged via digital means (Tian et al., 2020).

exceptional increase in governments' fiscal deficits and the consequential decrease in governmental funds available for incentivising the transition to more energy-efficient buildings (Tian et al., 2022).

Traditional funding mechanisms (e.g., government grants and incentives, loans) have demonstrated that they cannot cope with the growing demand for and need of capital to finance building renovation, so it is not surprising that the entire sector is constantly trying to attract more private investments (Tian et al., 2022). However, this is quite challenging given the scale of the investment required for these kinds of projects and the challenges associated with measuring the impact of "green" investments (United Nations, 2019).

Recent developments in the area of financial technologies (FinTech) have demonstrated how digital technologies can be leveraged to benefit both capital seekers (entrepreneurs, firms, and project promoters) and capital givers (investors), and therefore foster innovation in many sectors (Lynn & Rosati, 2021). FinTech can be seen as "a co-evolution and convergence of finance and technology" (Lynn et al., 2019, p. V) where new service providers typically leverage customer-centric platform-based business models enabled by the Internet and different degrees of disintermediation to overcome the limitations of the traditional financial system in terms of supply and access to capital, and the barriers to entry typical of traditional capital markets (Tönnissen et al., 2020; Lynn & Rosati, 2021; Sánchez, 2022). In so doing, they provide both small retail investors and large institutional investors with access to new investment opportunities, and capital seekers with additional funding they would not receive otherwise. In fact, in many cases, projects that seek funding through these alternative channels do not meet the requirements of traditional financial institutions in terms of credit history or are at an early stage of development, and therefore are not attractive to venture capitalist or investment funds. As such, these alternative sources of finance generate clear benefits not only for the parties involved in the transactions but for the economy as a whole (Sánchez, 2022).

While alternative sources of finance that are enabled by FinTech have gained significant traction in many sectors, the construction sector is still lagging behind in terms of adoption and is still mostly reliant on debt-based solutions (Ziegler et al., 2020). This suggests that FinTech solutions may play a pivotal role in supporting building renovation and therefore contributing to the ambitious sustainability targets that have been set by the EU and the United Nations (Economidou et al., 2019; United Nations, 2019).

In this chapter we focus on two main alternative sources of finance, namely crowdfunding, which is more established, and blockchain-based solutions, which are more novel and fast-growing.

10.3   Crowdfunding for Building Renovation

Economidou et al. (2011) provide an early identification of equity and debt crowdfunding as new and innovative sources of financing for building renovations. Crowdfunding is ideal in the manner in which it circumvents the constraints that exist in traditional bank financing, providing instead a new marketplace that allows for the pooling of financing from many retail investors ("the crowd") to support the building renovation project (Kunkel, 2015). Panteli et al. (2020) note that communities can become shareholders in energy efficiency projects through the mechanism of crowdfunding markets, allowing for greater buy-in from communities in the roll out of renewable energy and energy efficiency initiatives. Crowdfunding markets provide flexibility through connecting investors and beneficiaries directly, while offering lower costs of financing resulting from the use of the technology to facilitate the marketplace (Bertoldi et al., 2021). There are, of course, certain disadvantages to crowdfunding as articulated by Economidou et al. (2011): (a) the risk for beneficiaries in not securing the required level of funding, and (b) and the risk for investors in assuming all of the associated risks with extending the financing. It is within the latter context that the marketplaces for crowdfunding are less regulated than traditional markets.

In terms of project scale, Panteli et al. (2020) position crowdfunding as an ideal source of private financing for small-scale energy upgrading. Crowdfunding has the potential to form an important part of the funding mix, along with private-public and fully public funding mechanisms. An identified barrier to scaling up the amount of crowdfunding for building renovation is the wider public's understanding of crowdfunding markets.

Kunkel (2015) argues the merits for crowdfunding as a source of financing for building renovation as follows:

- The issue of information asymmetry is ideally mitigated through crowdfunding channels as crowdfunding platforms are effectively social networks that allow beneficiaries to engage directly with investors and reveal required information in an effective, low-cost manner. Indeed, beneficiaries often benefit from the collective expertise of the crowd, informing the project design, development, and implementation.
- Crowdfunding can also address concerns over implementation quality and trust in local partners and companies, as the crowd is likely to be local themselves and familiar with the parties. The crowd may be much better positioned than conventional banking institutions to appraise the implementation risks pertaining to the project in question.
- Finally, the local demographic of the crowd means that they are likely to benefit directly from the building renovation project beyond the

financial return and will be well placed to appraise the broader non-financial benefits, particularly in terms of the societal and environmental impact.

There are no studies, to the authors' knowledge, that empirically examine the crowdfunding of building renovation projects specifically. There is a more established literature however, albeit somewhat limited, that has studied the crowdfunding of real estate and renewable energy projects, which provides some useful insights.

In the context of real estate investment, Montgomery et al. (2018) use Disruptive Innovation Theory as a setting to appraise the potential for crowdfunding to be a disruptive source of financing. Based on a systematic literature review, the authors provide arguments for real estate crowdfunding as a disruptive innovation. Real estate crowdfunding is identified as offering cost and process efficiencies through technological innovation, having lower performance in certain areas (e.g., cybersecurity risk) relative to conventional financing channels, creating and facilitating a new marketplace for financing, and having less appeal among mainstream large real estate developers, while appealing to existing and new small- to medium-sized real estate developers. Shahrokhi and Parhizgari (2019) underscore the disruptive nature of real estate crowdfunding with a comparison against traditional financing, emphasising how the emergence of specialised crowdfunding platforms has overcome the high barriers historically to investment in real estate. Indeed, the authors note the explosion of platforms over recent years and the step change in real estate crowdfunding in the US from $1bn in 2009 to $17bn in 2015. Mamonov et al. (2017) confirm that real estate ventures are by the far the most successful proportion of the equity crowdfunding market in the US, constituting approximately 51% of all ventures that reached their minimum capital commitment target in the 2013–2016 period.

Through an empirical analysis of real estate crowdfunding campaigns in Italy, Gigante and Cozzio (2021) are able to identify the important determinants of successful crowdfunding campaigns, where success is defined as achieving (or exceeding) the target funding amount. Leveraging potential determinants from the general crowdfunding literature, the authors focus on the funding type (lending or equity), the duration of the investment, the minimum investment level for investors, and the expected annual return on investment. Duration is found to be important in that the longer the project the more difficult it is to secure the required funding. It is also found that higher expected returns attract investors and increase the chances of successfully securing the required funding. Borrero-Domínguez et al. (2020) conducted a similar study in the Spanish market. This study corroborates the findings of Gigante and Cozzio (2021) in showing that longer projects are less successful in securing funding, while projects that offer higher expected return are more successful. The authors also show that buy-to-sell projects are less successful than development loan projects, while greater levels of risk act as a deterrent for investors and impeding funding success.

In terms of the performance of real estate investment via crowdfunding markets, Schweizer and Zhou (2017) provide evidence that equity-based projects offer better returns, while higher levels of leverage are also associated with better returns. Other characteristics that lead to higher returns include provision for later payments to investors and higher minimum investment amounts.

In respect of the energy efficiency dimension to building renovation, it is worth exploring the literature that has examined the crowdfunding of renewable energy technology. Cumming et al. (2017), for instance, consider the determinants that drive crowdfunding. The authors show that price of oil is an important factor in determining the level of crowdfunding, with higher oil prices associated with a greater prevalence of crowdfunding directed at clean technology. The authors also show that the use of soft information (e.g., photos, video pitch, and text descriptions) is more prevalent in renewable energy-based crowdfunding campaigns and that this is used as a tactic to mitigate information asymmetry concerns for investors. It is shown further that the success of these crowdfunding campaigns is more sensitive to the use of soft information around the projects.

Slimane and Rousseau (2020), in a similar study, seek to identify the factors that can lead to a successful crowdfunding campaign. Financial characteristics of the renewable energy project are found to be important, including the interest rate applied, the funding amount requested, the size of the firm in question, and the overall financial performance of the firm. Non-financial characteristics such as age and gender of the entrepreneur, in addition to the size of their social network, are found to be relevant.

While such studies demonstrate that crowdfunding can be successful from the beneficiaries' perspective, what is the impact of such crowdfunding? Appiah-Otoo et al. (2022) provide evidence to support the tangible impact that crowdfunding can have on renewable energy development. The authors demonstrate that on a cross-country basis a 1% increase in crowdfunding bolsters actual renewable energy generation by 0.35%. Indeed, the authors further show a very interesting bi-directional causal relationship between crowdfunding and renewable energy generation. This suggests that the development of crowdfunding markets helps to channel the financing to expand renewable energy generation, while the expansion of renewable energy generation helps to attract investors to crowdfunding markets who are looking for investment opportunities.

Of course, the above insights are on the demand side of crowdfunding (i.e., the beneficiaries). One also needs to consider the supply side of crowdfunding (i.e., the investors). Understanding investor perceptions and behaviours is pivotal here. Bergmann et al.'s (2021) study, for example, is one such study that provides qualitative cross-country survey evidence that a significant majority of those surveyed have a strong awareness of the existence of crowdfunding markets, with almost half having invested in such marketplaces previously. A significant minority (~40%) indicated an intention to invest in renewable energy projects through crowdfunding channels over the next three years.

Literature also tells us that the platform has a central role to play in the successful mobilisation of crowdfunding to renewable energy projects. For example, De Broeck (2018) studies best practices in respect of platforms servicing investment in renewable energy projects. The qualitative analysis provides insights across a number of key dimensions of crowdfunding activity around renewable energy projects: the impact of regulation, risk exposures resulting from the underlying platform business models, and the platforms' attitude towards risk.

De Broeck (2018) finds that crowdfunding activity around renewable energy projects is strongest in jurisdictions where there is strong policy support for renewable energy, citing premium tariffs and/or feed-in-tariffs, which offer better long-term certainty over the cash flows associated with the renewable energy projects. When assessing the platforms on the basis of credit risk, De Broeck (2018) is able to identify a set of platforms that work to a combination of low risk supports (such as feed-in-tariffs) and low risk instruments (secured business loans, bonds/debentures, and senior bond loans), while another set of platforms works to a combination of very low risk tariff premiums and high risk instruments (subordinate profit participating loans). The presence of strong support is seen as an important measure for the mitigation of credit risk for investors, which encourages more crowdfunding activities. De Broeck (2018) also finds that due diligence procedures are deemed to be the most significant measure that platforms can take to mitigate the credit risk exposure of investors. Platforms that reduce credit risk exposure ensure greater and more persistent levels of engagement from investors, protecting the resulting supply of funding to renewable energy projects.

10.4   BLOCKCHAIN FOR BUILDING RENOVATION

Blockchain technology was originally proposed in 2008 by Satoshi Nakamoto as the technology underpinning Bitcoin (Nakamoto, 2008). While most of the attention around blockchain was initially devoted to payment and other transactional systems, a number of alternative use cases across different industries have emerged over time. With a specific focus on the built environment, for instance, Arup (2019) considers blockchain applications in the context of property, but also the wider and associated areas of smart cities, energy, transport, and water. Khatoon et al. (2019) note how blockchain is being considered in areas such as large-scale energy trading systems, peer-to-peer energy trading, project financing, supply chain tracking, and asset management. The focus of Khatoon et al. (2019) is on the application of blockchain in energy efficiency, where they show that blockchain-based smart contracting provides a solution to efficient and transparent trading of energy efficiency savings. Blockchain also offers potential for efficient building information management, with Liu et al. (2021) reviewing the literature towards addressing gaps in the smart city context.[4] Woo et al. (2021) provide a similar review with specific focus on building energy management. The remainder of this section focuses on three areas—energy performance contracting, building and renovation financing, and digital twinning.

We begin with energy performance contracting. An energy performance contract (EPC) is described as a creative financing mechanism that funds energy upgrades in, for example, building renovation works.[5] The EPC involves a contract with an assigned energy services company (ESCO) that designs and delivers on the energy efficiency plan, with the (future) revenues from the resulting costs savings being used to net off against the

---

[4] Relatedly, there is a literature that has considered the role that the Internet of Things can play in the real-time monitoring and management of building information. See, for example, Altohami et al. (2021) for a review.

[5] https://e3p.jrc.ec.europa.eu/articles/energy-performance-contracting#:~:text=Energy%20Performance%20Contracting%20(EPC)%20is,energy%20upgrades%20from%20cost%20reductions.

(primarily upfront) expenses around the project. Aoun (2020) notes that EPCs are suitable when funding sources are elusive, maintenance is lacking, or new equipment and technology is needed and requires unique skills. The EPC area has been well studied for a considerable period of time; Zhang and Yuan (2019) provide a comprehensive review of recent literature.

Blockchain is of interest in the area of energy performance contracting as the technology offers scope to introduce efficiencies into the process, while it also allows for trust to be established between the parties involved in the building renovation given the integrity of the blockchain. Schletz et al. (2020), for example, discuss how blockchain can provide an alternative channel through which to raise the required capital for the energy efficiency plan underlying an EPC. This utilises the process of tokenisation. Engineering digital tokens for sale to investors over a blockchain allows a way to pool funding from a large array of both retail and institutional investors. This is effectively a crowdfunding market, akin to what we met previously, but rather than being based on traditional debt and equity instruments, it is based on digital tokens[6] and fully decentralised. Schletz et al. (2020) propose the use of security tokens—which are more strongly regulated versions of digital tokens and which may reflect more closely traditional debt and equity instruments—under such blockchain applications.[7] Blockchain-based smart contracts then allow the automated transfer of the capital raised to the ESCO, while it also allows for income, as defined under the security token specification, to be transferred back to the investors. Aoun (2020) provides a wider discussion, proposing a blockchain model design suitable for energy performance contracting, which builds trust for the main players involved: customers, investors, and the ESCO. Exploitation of smart contracts is proposed for (1) the efficient recording of data collected from the implemented energy conservation measures, specifically logging data (via oracle technology) from external sensors in a smart contract (data logger smart contract); (2) the calculation of the daily adjusted baseline energy consumption based on the logged data and some agreed formulation, and the calculation of the

---

[6] While a discussion of digital tokens is beyond the scope of this chapter, the interested reader is directed to, for example, Tasca (2019) for a review of token-based business models.

[7] Stekli and Cali (2020) also consider the potential of security tokens as an equity crowdfunding channel for offshore wind energy, while Halden et al. (2021) do similarly for solar energy.

actual daily savings achieved with reference to this baseline (adjustments smart contract); and (3) the incrementing of the monthly savings record with the calculated daily savings (savings smart contract). Gürcan et al. (2018) similarly consider how blockchain can potentially reconcile, in the case of energy performance contracting, the requirement to process and analyse large volumes of data and the requirement to implement complex algorithms to determine the baseline energy consumption against which the actual energy consumption is benchmarked.

Blockchain can, more generally, facilitate funding release in the real estate market. While the concept of real estate tokenisation is new, the market is developing and use cases are emerging. A widely referenced case is AspenCoin, the first real estate Security Token Offering (STO). Launched in 2018, it raised US $18 million within a 2-month period in exchange for 18.9% of the ownership of the St. Regis Aspen Resort in Aspen, Colorado (Carroll, 2018). Real estate tokenisation offers fractional ownership opportunities, widening the funding pool for real estate investments and creating liquid secondary real estate markets where the trading of real estate tokens can occur (Baum, 2021). In the context of commercial real estate, Smith et al. (2019) also emphasise the benefits of blockchain in terms of securitisation and trading, but extend the discussion to the potential application of blockchain to the real estate investment value chain and to the representation of the physical assets. Smart contracts are again core to these blockchain applications allowing for automation of processes. From an empirical perspective, Swinkels (2022) provides one of the first studies of the real estate token market in the US, providing evidence that tokenisation is indeed leading to notable fractionalisation of ownership. Furthermore, Swinkels (2022) documents an alignment between the prices of real estate tokens and the US house price index, showing an integration of virtual and real property markets.

Finally, blockchain has considerable potential in the area of digital twinning. Hunhevicz et al. (2022) consider how blockchain can be integrated and exploited leveraging a blockchain-based business model that relies on interaction between the physical building environment and the virtual building environment. The latter serves to simplify the connection between the real world data and the smart contracts, reducing the data storage requirement of the smart contract. Similar to the previous studies, the blockchain is shown to be useful in delivering funding into the building project via digital tokenisation, and in the automated execution of the main phases of the energy performance contract via a smart contract, while it further allows for trust in the transactions between all parties involved.

10.5 CONCLUSION

This chapter summarises the somewhat limited literature that exists addressing the intersection of the financial technology and the building renovation domains. This deficit of knowledge means that there is a tangible opportunity to advance research in the directions outlined in respect of non-blockchain-based crowdfunding and blockchain-based crowdfunding, although the latter will take some years for the required token-based marketplaces to emerge and mature. Given the EU's present focus on overhauling the existing Energy Efficiency Directive towards achieving its ambitious building renovation targets, the potential for meaningful policy impact from timely research is pronounced.

From our discussion of non-blockchain-based crowdfunding, it is evident that there is a deficit of knowledge and empirical evidence in respect of the crowdfunding of building renovation. Little is known on the demand side (crowdfunding beneficiaries) or the supply side (crowdfunding investors), or indeed on the responsibilities of crowdfunding platforms. The existing literature on crowdfunding for real estate investments and renewable energy projects literature provides some useful insights that are likely to be relevant in the building renovation space. However, dedicated empirical studies that track the crowdfunding directed at building renovation projects are required, while an understanding of whether and how crowdfunding platforms promote and support building renovation projects (relative to new building development projects) is needed in order to assess the funding landscape holistically in the context of the built environment. More insight is also required into customer views of crowdfunding as a channel to finance building renovation. There are idiosyncratic features to building renovation that require more thoughtful consideration to appraise how crowdfunding can be optimised to deliver on the required scale of building renovation. In the case of the EU, such tailored research would have the potential to impact building renovation policy.

In respect of blockchain-based crowdfunding, the nascent nature of these market innovations means that time will reveal much information on the success of such blockchain applications. Future studies may attempt to answer the question: how can tokenisation most effectively work as a funding release mechanism (beyond energy performance contracting) for building renovation specifically? Our exploration of blockchain in respect of energy performance contracting is clearly new and the literature sparse. As technical blockchain developments continue in practice, we will likely see the emergence of active token-based markets that will drive funding towards building renovation work. Similar to the knowledge gaps identified in previous sections, empirical evidence will need to be accumulated in respect of the demand side (beneficiaries) and the supply side (investors) of these token-based markets. What drives a successful Security Token Offering will be important to ascertain, while the comparison of such blockchain-based crowdfunding will need to be compared against existing non-blockchain-based equity and debt crowdfunding. Furthermore, as we see greater adoption of smart contracts in energy performance contracting, we will be able to appraise the effectiveness of the financing mechanism in terms of its return performance and risk profile.

References

Altohami, A. B. A., Haron, N. A., Ales, A. H., & Law, T. H. (2021). Investigating approaches of integrating BIM, IoT, and facility management for renovating existing buildings: A review. *Sustainability, 13*(7), 3930.

Aoun, A. G. (2020). Blockchain application in energy performance contracting. *International Journal of Strategic Energy & Environmental Planning, 2*(2).

Appiah-Otoo, I., Song, N., Acheampong, A. O., & Yao, X. (2022). Crowdfunding and renewable energy development: What does the data say? *International Journal of Energy Research, 46*(2), 1837–1852.

Arup (2019). *Blockchain and the Built Environment.* https://www.arup.com/-/media/arup/files/publications/b/blockchain-and-the-built-environment.pdf

Baum, A. (2021). Tokenization—The future of real estate investment? *The Journal of Portfolio Management, 47*(10), 41–61.

Beck, R., Avital, M., Rossi, M., & Thatcher, J. B. (2017). Blockchain technology in business and information systems research. *Business & Information Systems Engineering, 59*(6), 381–384.

Belleflamme, P., Lambert, T., & Schwienbacher, A. (2014). Crowdfunding: Tapping the right crowd. *Journal of Business Venturing, 29*(5), 585–609.

Bergmann, A., Burton, B., & Klaes, M. (2021). European perceptions on crowdfunding for renewables: Positivity and pragmatism. *Ecological Economics, 179*, 106852.

Bertoldi, P., Economidou, M., Palermo, V., Boza-Kiss, B., & Todeschi, V. (2021). How to finance energy renovation of residential buildings: Review of current and emerging financing instruments in the EU. *Wiley Interdisciplinary Reviews: Energy and Environment, 10*(1), e384.

Borrero-Domínguez, C., Cordón-Lagares, E., & Hernández-Garrido, R. (2020). Sustainability and real estate crowdfunding: Success factors. *Sustainability, 12*(12), 5136.

Brown, D., Hall, S., Martiskainen, M., & Davis, M. E. (2022). Conceptualising domestic energy service business models: A typology and policy recommendations. *Energy Policy, 161*, 112704.

Carroll, R. (2018). In $18 million deal, nearly one-fifth of St. Regis Aspen sells through digital tokens. *Aspen Times.* Retrieved November 7, 2022, from https://www.aspentimes.com/trending/in-18-million-deal-nearly-one-fifth-of-st-regis-aspen-sells-through-digital-tokens/

Cumming, D. J., Leboeuf, G., & Schwienbacher, A. (2017). Crowdfunding cleantech. *Energy Economics, 65*, 292–303.

De Broeck, W. (2018). Crowdfunding platforms for renewable energy investments: An overview of best practices in the EU. *International Journal of Sustainable Energy Planning and Management, 15*, 3–10.

Economidou, M., Todeschi, V., & Bertoldi, P. (2019). *Accelerating energy renovation investments in buildings.* Publications Office of the European Union.

Economidou, M., et al. (2011). *Europe's buildings under the microscope. A country-by-country review of the energy performance of buildings.* Buildings Performance Institute Europe (BPIE). https://bpie.eu/wp-content/uploads/2015/10/HR_EU_B_under_microscope_study.pdf

Eyre, N. (2013). Energy saving in energy market reform—The feed-in tariffs option. *Energy Policy, 52*, 190–198.

Gigante, G., & Cozzio, G. (2021). Equity crowdfunding: An empirical investigation of success factors in real estate crowdfunding. *Journal of Property Investment & Finance.*

Glaser, F. (2017). *Pervasive decentralisation of digital infrastructures: A framework for blockchain enabled system and use case analysis.* 50th Hawaii International Conference on System Sciences (HICSS 2017), Waikoloa, HI, USA.

Gürcan, Ö., Agenis-Nevers, M., Batany, Y. M., Elmtiri, M., Le Fevre, F., & Tucci-Piergiovanni, S. (2018, June). An industrial prototype of trusted energy performance contracts using blockchain technologies. In *2018 IEEE 20th International Conference on High Performance Computing and Communications; IEEE 16th International Conference on Smart City; IEEE 4th International Conference on Data Science and Systems (HPCC/SmartCity/DSS)* (pp. 1336–1343). IEEE.

Halden, U., Cali, U., Dynge, M. F., Stekli, J., & Bai, L. (2021). DLT-based equity crowdfunding on the techno-economic feasibility of solar energy investments. *Solar Energy, 227*, 137–150.

Hunhevicz, J. J., Motie, M., & Hall, D. M. (2022). Digital building twins and blockchain for performance-based (smart) contracts. *Automation in Construction, 133,* 103981.

Kaplow, S. D. (2008). Does a green building need a green lease. *The University of Baltimore Law Review, 38,* 375.

Karakosta, C., Papapostolou, A., Vasileiou, G., & Psarras, J. (2021). Financial schemes for energy efficiency projects: Lessons learnt from in-country demonstrations. In *Energy services fundamentals and financing* (pp. 55–78). Academic Press.

Khatoon, A., Verma, P., Southernwood, J., Massey, B., & Corcoran, P. (2019). Blockchain in energy efficiency: Potential applications and benefits. *Energies, 12*(17), 3317.

Kunkel, S. (2015). Green crowdfunding: A future-proof tool to reach scale and deep renovation? In *World sustainable energy days next 2014* (pp. 79–85). Springer.

Lee, P., Lam, P. T. I., & Lee, W. L. (2015). Risks in Energy Performance Contracting (EPC) projects. *Energy and Buildings, 92,* 116–127.

Liu, Z., Chi, Z., Osmani, M., & Demian, P. (2021). Blockchain and Building Information Management (BIM) for sustainable building development within the context of smart cities. *Sustainability, 13*(4), 2090.

Lukas, A. A. (2018). *Financing energy efficiency, part 1: Revolving funds* (Vol. no. 129733, pp. 1–12). The World Bank.

Lynn, T., & Rosati, P. (2021). New sources of entrepreneurial finance. *Digital Entrepreneurship, 209.*

Lynn, T., John G. M., Pierangelo R., & Mark C. (2019). *Disrupting Finance: FinTech and Strategy in the 21st Century.* Springer Nature.

Mamonov, S., Malaga, R., & Rosenblum, J. (2017). An exploratory analysis of Title II equity crowdfunding success. *Venture Capital, 19*(3), 239–256.

Montgomery, N., Squires, G., & Syed, I. (2018). *Disruptive potential of real estate crowdfunding in the real estate project finance industry: A literature review.* Property Management.

Nakamoto, S. (2008). Bitcoin: A peer-to-peer electronic cash system. *Decentralized Business Review, 21260.*

O'Shields, R. (2017). Smart contracts: Legal agreements for the blockchain. *North Carolina Banking Institute, 21,* 177.

Oracle. (2016). *Leasing—An overview.* Oracle Corporation. https://docs.oracle.com/cd/E74659_01/html/LE/LE02_Intro.htm

Panteli, C., Klumbytė, E., Apanavičienė, R., & Fokaides, P. A. (2020). An overview of the existing schemes and research trends in financing the energy upgrade of buildings in Europe. *Journal of Sustainable Architecture and Civil Engineering, 27*(2), 53–62.

Rosati, P., & Čuk, T. (2019). Blockchain beyond cryptocurrencies. *Disrupting Finance,* 149–170.

Sánchez, M. A. (2022). A multi-level perspective on financial technology transitions. *Technological Forecasting and Social Change, 181*, 121766.

Schletz, M., Cardoso, A., Prata Dias, G., & Salomo, S. (2020). How can blockchain technology accelerate energy efficiency interventions? A use case comparison. *Energies, 13*(22), 5869.

Schweizer, D., & Zhou, T. (2017). Do principles pay in real estate crowdfunding? *The Journal of Portfolio Management, 43*(6), 120–137.

Shahrokhi, M., & Parhizgari, A. M. (2019). Crowdfunding in real estate: Evolutionary and disruptive. *Managerial Finance, 46*(6), 785–801.

Slimane, F. B., & Rousseau, A. (2020). Crowdlending campaigns for renewable energy: Success factors. *Journal of Cleaner Production, 249*, 119330.

Smith, J., Vora, M., Benedetti, H., Yoshida, K., & Vogel, Z. (2019). Tokenized securities and commercial real estate. SSRN: 3438286.

Stekli, J., & Cali, U. (2020). Potential impacts of blockchain based equity crowdfunding on the economic feasibility of offshore wind energy investments. *Journal of Renewable and Sustainable Energy, 12*(5), 053307.

Swinkels, L. (2022). Empirical evidence on the ownership and liquidity of real estate tokens. SSRN: 3968235.

Tasca, P. (2019). Token-based business models. In *Disrupting finance* (pp. 135–148). Palgrave Pivot.

Tian, J., Yu, L., Xue, R., Zhuang, S., & Shan, Y. (2022). Global low-carbon energy transition in the post-COVID-19 era. *Applied Energy, 307*, 118205.

Tian, Y., Lu, Z., Adriaens, P., Minchin, R. E., Caithness, A., & Woo, J. (2020). Finance infrastructure through blockchain-based tokenization. *Frontiers of Engineering Management, 7*(4), 485–499.

Tönnissen, S., Beinke, J. H., & Teuteberg, F. (2020). Understanding token-based ecosystems—A taxonomy of blockchain-based business models of start-ups. *Electronic Markets, 30*(2), 307–323.

United Nations. (2019). *Harnessing digitalization in financing of the sustainable development goals*. United Nations. https://unsdg.un.org/sites/default/files/2020-08/DF-Task-Force-Full-Report-Aug-2020-1.pdf

Woo, J., Fatima, R., Kibert, C. J., Newman, R. E., Tian, Y., & Srinivasan, R. S. (2021). Applying blockchain technology for building energy performance measurement, reporting, and verification (MRV) and the carbon credit market: A review of the literature. *Building and Environment, 205*, 108199.

Zhang, W., & Yuan, H. (2019). A bibliometric analysis of energy performance contracting research from 2008 to 2018. *Sustainability, 11*(13), 3548.

Ziegler, T., et al. (2020). *The global alternative finance market benchmarking*. Cambridge Centre for Alternative Finance. https://www.jbs.cam.ac.uk/wp-content/uploads/2020/08/2020-04-22-ccaf-global-alternative-finance-market-benchmarking-report.pdf

Index[1]

---

[1] Note: Page numbers followed by 'n' refer to notes.

© The Author(s) 2023

T. Lynn et al. (eds.), *Disrupting Buildings*, Palgrave Studies in Digital Business & Enabling Technologies, https://doi.org/10.1007/978-3-031-32309-6